DA 86.22 .D7 C8

Corbett, Julian Stafford, Sir, 1854-1922.

Sir Francis Drake

English Men of Action

SIR FRANCIS DRAKE

SIR FRANCIS DRAKE

SIR FRANCIS DRAKE

BY

JULIAN CORBETT

GREENWOOD PRESS, PUBLISHERS
WESTPORT, CONNECTICUT

DA
86.22
.D7
C8

Originally published in 1890
by Macmillan and Co., London and New York

First Greenwood Reprinting 1970

Library of Congress Catalogue Card Number 69-13865

SBN 8371-4086-2

Printed in the United States of America

HOFSTRA UNIVERSITY
LIBRARY

CONTENTS

CHAPTER I

THE REFORMATION MAN

OF all the heroes whose exploits have set our history aglow with romance there is not one who so soon passed into legend as Francis Drake. He was not dead before his life became a fairy tale, and he himself as indistinct as Sir Guy of Warwick or Croquemitaine. His exploits loomed in mythical extravagance through the mists in which, for high reasons of State, they long remained enveloped, and to the people he seemed some boisterous hero of a folk-tale outwitting and belabouring a clumsy ogre.

> And that our Drake might David parallel,
> A *mass* of Man, a gyant he did quell.

So punned a west-country Protestant; and even now the most chastened explorer of pay-sheets and reports cannot save his imagination from the taint of the same irrational exultation that possessed the Admiral's contemporaries. The soberest chroniclers reeled with unscholarly gait as they told the tale, and the most dignified historians made pedantic apology for the capers they felt forced to cut. From his cradle to his grave the story is one long draught of strong waters, and the very first sip intoxicates. Peer into the mists that fitly shroud his birth and all is dark, till on a sudden the veil is riven with an outburst of Catholic fury. Then, while the flash of the explosion illuminates the scene, a small party of desperate Protestants are seen flying for their lives, and in their midst a blue-eyed, curly-haired child, scarce out of babyhood, who is Francis Drake.

So Reformation set her seal in his forehead at the outset. It was in the year 1549, when Edward the Sixth was king, and on Whitsunday the new service-book was to be read for the first time throughout the realm. To the fervent simplicity of the west-country folk, to whom the mass was the beginning and the end of religion, it was as though Christ were being banished from the earth, and ere the week was out all Devon and Cornwall were in a blaze of religious riot. In the heart of the conflagration lay Tavistock, where still green memories of the kindly monks added fuel to the flames. Little mercy was there in the shadow of the old abbey walls for active partisans of the new order. About the great centres of trade there was now growing up on the ruins of the Middle Ages a party democratic in politics and religion, the nucleus of the revolutions to come, and of such was little Francis's father, Edmund Drake. He had once been a sailor they say, and that is not unlikely. For his kinsman, old William Hawkins, like his father before him, was a great merchant and shipowner of Plymouth, and, first of all Englishmen, had sailed to the Brazils in King Henry's time. Now, however, Edmund Drake had taken his place among the lesser western gentry, and was settled down in substantial comfort at Crowndale, hard by the town of Tavistock.[1] There he had won himself powerful friends, as a strong "Reformation man" with a turn for preaching, which in those days, when politics and religion were not yet divorced, took the place of political speaking. The great Earl of Bedford, himself the most powerful of the Protestant leaders, bestowed upon him his patronage. The Earl's eldest son, Francis Russell, held the preacher's first-born at the font, and endowed him with his own name, as he afterwards endowed Francis Bacon. Thus honourably the flail of the Papacy was baptized into the Protestant faith; but now the preacher's great friends were only a source of danger. There could be for him no thought but flight. The most powerful of his political patrons could not shield him where he was; for the Earl himself, with all the forces he could muster at his back, dared not approach within fifty miles of his own seat at Tavistock. But in the good Protestant town of Plymouth Edmund Drake had friends to shelter him, for William Hawkins and his sons owned a great part of the town. Out in the harbour lay St. Nicholas island, which in the years to come was to be honoured with the blue-eyed baby's

[1] The idea that Drake was of low birth seems to have arisen from a misapprehension of the word "mean," by which contemporary translators rendered Camden's *mediocri loco.* We should now write "moderate birth" or "middle class." The patient investigations of Dr. Drake and others have shown clearly that the Drakes of Tavistock freely intermarried with the lesser Devon gentry, and Francis Drake not only seems to have borne arms before the Queen's grant (Peralta, *Costa Rica*, etc., p. 583), but was in the habit of using those she gave him as an augmentation upon his family coat.

name, and there, as a throng of fugitives gathers for sanctuary, darkness falls upon the preacher's flight.

But it is only to startle us again out of all sobriety when next the veil is lifted, so like a fairy tale the truth appears. In Chatham reach, off the new dockyard, was the anchorage where the navy ships were laid up when out of commission, and there too lay veteran war-hulks slowly rotting to death. So well had Edmund Drake's friends stood by him that one of these had been assigned to him as a dwelling-place, and with it an official appointment as Reader of prayers to the Royal Navy. To such a nursery had Catholic devotion driven the most redoubtable of its enemies. What wonder that it bred a crusading sea-king! The clatter of the ship-wrights' hammers in the dockyard, the sea-songs of the mariners as they polished the idle guns, the fierce and intemperate denunciations of his father's friends vowing vengeance on the "idolaters who had defiled the House of God,"—such were the first sounds his dawning intelligence learnt to grasp. His eyes could rest nowhere but on masts, and guns, and the towering hulks of the warships which lay anchored about his floating home. His very play-things were instruments of destruction; the prayer he lisped at his mother's knee was little better than a curse.

So passed the first years of his boyhood, and year after year was born another sturdy little Protestant till Edmund Drake had round him twelve young champions of his hot opinions. "As it pleased God," the old chronicler rejoiced to say, "to give most of them a being on the water, so the greatest part of them died at sea." Boys whose lullaby had been the rush of the tide and the hum of the wind in the standing rigging were marked by destiny for a sailor's life, and the influence which their father commanded seemed to open the navy to their ambition. But as Francis approached the age of apprenticeship all his interest was lost at a stroke. In the summer of 1553 the sickly young king breathed his last, and a Catholic princess reigned in his stead. Drake's party found itself fallen from the Delectable Mountains of Patronage into the Valley of the Shadow of Death, and soon Protestant England was chafing ominously at the news that Mary was to marry with the Prince of Spain. The new faith, the very independence of England seemed at stake, and it was under young Drake's eyes that the storm of opposition burst. He must have seen Wyatt ride into Rochester and establish his headquarters in the castle. He must have heard him call on all true Englishmen to rally to his standard to save the country from the Pope and Spain. He must have seen the fleet in the Medway supplying the patriot rebel with artillery, and shouted with the rest to see the Duke of Norfolk recoil before Wyatt's banner from Rochester bridge. Then came the pause while London was beleaguered, and then the block and gibbet were busy with those who had failed. Friends, namesakes, perhaps even kinsmen of the Drakes, suffered with poor Wyatt, and Francis was at least old enough to know it was because they had lifted their hands against Spain and Rome. For the issue was so clear, and feeling so intense, that children forgot their games to play at politics. They snowballed the suite of the Spanish Ambassador, they fought mock combats between Wyatt and the Prince of Spain, and once were barely prevented from hanging the lad who represented Philip.

These were the boy's first recollections, and upon them came a rude change of fortune to heap up the hate that was gathering in his masterful temper for Rome and Spain. The preacher's occupation was gone, his prospects shattered, and he had to seize any opportunity to launch his sons into the world. Francis was apprenticed to the skipper of a small craft that traded to France and Holland. It was a poor end to his brighter expectations. The hardships of a ship-boy on board a Channel coaster in those days are to us inconceivable. In danger, privation, and exposure, the lad was moulded into the man, and even as his frame was being rudely forged into the thick-set solidity that distinguished his manhood, so was his spirit being tempered in the subtlest medium that destiny could have chosen. As he passed to and fro upon the narrow seas in the months of his hottest youth, he was plunged into the most violent religious passion which the Reformation ever evoked. For ere he was well on the threshold of manhood, Philip was goading his Low Country subjects into a frenzy with his insane persecutions. On quay, and market, and shipboard the horror of the Inquisition was the only talk, and the Flemings were flying for sanctuary to England. Elizabeth, who had now begun her reign, received them with open arms, and the preacher too held up his head as the tide turned once more. His Devonshire friends and patrons were those who had stood most stoutly by the young princess in the darkest hours of her danger. They were now all-powerful, and Edmund Drake was gladdened with the living of Upchurch on the Medway. Fortune smiled on Francis no less. His master died, and out of love for the lad who had served him so well left him the vessel on which he had been apprenticed. The young skipper could thus begin to trade on his own account; and it can hardly have been but that he brought over bands of Flemish refugees, and caught from them something of their defiant and implacable attitude towards their persecutor.

Year by year the grumbling of the coming storm grew louder, and the narrow seas began to swarm with Protestant rovers revenging themselves with wanton cruelty upon Catholic ships. England was their base and market, and at last, in January, 1564, Spain, in a fit of just exasperation, closed her ports and seized every English vessel on which she could lay her hands. Drake's trade was stopped, but it mattered little. He sold his vessel and entered the service of his two kinsmen, old William Hawkins's adventurous sons. A wiser step he could not have taken. The brothers, already large ship-owners at Plymouth and London, were more than maintaining the family name for skill and enterprise. Captain John, the younger brother, had just returned triumphant from that first slaving voyage of his which so darkly ushered in the grandest era of English maritime adventure. The shareholders were revelling in an unheard-of profit, and court, commerce, and admiralty were bowing before the brothers as society now caresses the last enthroned financial king. In October, 1564, John Hawkins sailed again to repeat his happy venture, but Drake did not accompany him. As soon as diplomacy had removed the embargo he had sailed as purser of a ship, belonging probably to William Hawkins, to the Biscayan province of Spain, and once more it seems as though the finger of Destiny had beckoned him there to show the work he was born to do. St. Sebastian was the chief port of Biscaya, and there at this moment were creeping from the pestilential dungeons of the Inquisition the remnants of a Plymouth crew, who had been seized when the embargo was first proclaimed. In six months half of them had rotted to death, and it may even have been that his ship brought home the broken wretches that survived.

So successful was John Hawkins's second voyage, and so alarming the activity it bred in the English ports, that Spain began to tremble for her monopoly of the western trade. She had absolutely forbidden her American subjects to traffic with foreigners, and particularly in negro slaves, and so indignantly did the Ambassador protest against Hawkins's conduct, that the Council, still ignorant of their strength, felt themselves obliged to bind him over the following year not to go to the Indies. But if he did not go, an expedition went. It was under the command of a Captain Lovell, one of the forgotten pioneers of North America, and with it sailed Francis Drake. It was his first sight of the fabled Indies, and one he never forgot. For in attempting to set the prohibition at defiance in the port of La Hacha, on the Spanish Main, they found themselves the victims of some treacherous stratagem which sent them home with the loss of all their venture.

It was a blow Drake never forgot nor forgave, but in the following year the attempt was not repeated, and he sought to recoup his shattered fortunes by serving in a voyage to Guinea. It was probably that under Captain George Fenner; and, if so, he must have witnessed that brilliant engagement, in which for two days with his own single ship and a pinnace Fenner fought and finally drove off a great Portuguese galleasse and six gunboats. It was the first action of a long and glorious series, and the news of it came most timely to add its inch to the lengthening stride of the epic. For the Netherlands were sullenly turning upon their Spanish governor, the English Catholics were staring dumbfoundered at the blackened relics of Darnley's murder, and Elizabeth felt she could for the present snap her fingers at the Spanish Ambassador and indulge in a little more buccaneering.

It was her favourite investment. For her the risk was small and the hopes of profit too rosy to be resisted. It seems strange conduct for a great Queen, but she had to encourage adventurous commerce, on which, in those days of a half-established navy, England's maritime position depended. The royal ships were merely a nucleus round which armed merchantmen gathered in time of war. It was as natural for the Queen to employ her ships in commerce while the realm was at peace, as it was for shipowners to accept a charter-party from the admiralty at the outbreak of a war. The mercantile marine then formed what we should now call the naval reserve. The situation was perfectly understood and recognised by both Government and shipowners. Private cruisers were a necessity to every considerable owner. He kept them, as large firms now insure their own ships; and at a time when the diplomatic system was not yet established, a merchant who considered himself injured abroad had more faith in reprisals with his cruisers than in complaints to his Government.

In such a state of things it is hardly to be wondered at that the line was not always very sharply defined between naval and commercial expeditions. In the present case there is little doubt that both the Hawkinses and Elizabeth had scores to settle in connection with the La Hacha affair, and the rough usage of the last expedition to Guinea. The Queen's name, of course, did not appear. It never did. It was nominally a venture by Sir William Garrard and Co., in which the Hawkinses were the largest subscribers. The Queen's contribution was two ships of war. This was her usual practice. They cost her nothing. They had merely to be valued—not often, it would seem, much below their worth—and Her Majesty then stood as a shareholder to the extent of the valuation. Not a penny of cash was she wont to provide. The Company had even to fit out the ships for sea. She had but little to lose and everything to gain, and the temptation to filibuster under such terms is not difficult to appreciate.

Such was the expedition which on October 2nd, 1567, sailed out of Plymouth harbour with John Hawkins as admiral, and Francis Drake as pilot or second officer of his ship.[1] It consisted of the *Jesus* and the *Minion* of Her Majesty's navy, and four other vessels which the Company had chartered of the Hawkinses. In no way did it differ from a naval squadron. It had its admiral, its vice-admiral, and its captain of the land forces. It had every kind of munition of the latest type; it even carried field-artillery, and its crews had been completed by the pressgang. The first rendezvous was fixed at the Canaries, and thence early in November the squadron sailed for the west coast of Africa. They were

[1] *Memorias de los Corsarios Ingleses . . . en las Indias*, S. P. Spain, 1580, bundle xviii.

now well within the Portuguese sphere of action, and no time was lost in exacting reprisals for Fenner's ill-usage. Trade in these regions was carried on in vessels called caravels. They were rigged and fitted like galleys, with a lofty square poop, and being of light draught, they were admirably adapted for entering the rivers and inlets where the trade was done. One of these was picked up before the squadron reached Cape Blanc, and on the way to Cape Verde another was sighted. It had been captured by a Frenchman, but this made no difference to Hawkins. The *Minion* gave chase, and took it without compunction. It proved to be a smart new craft of one hundred and fifty tons, and as two pinnaces had been lost in the foul weather that had prevailed, it was permanently attached to the squadron, and Francis Drake placed in command.

For three months the squadron continued on the coast hunting for negroes and Portuguese caravels, and Drake, in the *Grace of God*, was not behindhand in landing and burning and cutting out. It was work he could enjoy without compunction, though he was as religious as Hawkins himself and quite as humane. The institution of the slave-trade was the first genuine attempt at the abolition of slavery. Las Casas himself, the apostle of the Indies, the father of philanthropy, had been its ardent advocate. Forced labour in the American mines and plantations was rapidly exterminating the natives. By importing black labour from the pestilential heathendom of Africa to the Christian paradise in the west, the saintly missionary thought not only to confer a temporal and spiritual blessing upon the negroes themselves, but also to save the

Indians without ruining the colonists. So fairly did the idea promise, that it seemed an inspiration from Heaven. Its evils, of course, soon pronounced themselves, and Philip had forbidden the trade except under special license from himself. Of this the English understood nothing; and the old Puritan captains went on hunting slaves, just as they prayed and fought, with all their heart and with all their strength, and never knew a reason why they should not.

By the end of January some five hundred negroes had been collected, and the squadron sailed for the Spanish Main. The French captain seems to have been persuaded to join hands with Hawkins, for Drake was transferred to the *Judith*, a barque of fifty tons and one of the original squadron. In seven weeks they were lying off the island of Margarita. It was the depot from which were supplied the struggling settlements on the Spanish Main, as the north coast of South America was then called, and here in spite of the Spaniards' protests the fleet quietly revictualled. It had now been five months at sea, and in those days ships' bottoms grew so quickly foul that it was already necessary to clean them. With the same effrontery which marked their dealings at Margarita, the well-stored squadron put into a lonely little port somewhere between Caracas and Coro in the Golfe Triste, and there for two months they stayed, leisurely careening, scraping, and refitting their weather-beaten ships.

Then trade began in earnest, and as lawfully as might be. It is a story that has been told more than once in the glorious and disreputable annals of British enterprise, and not so long ago about opium on the coast of

China. The Spaniards of course refused to buy negroes, as the Chinese refused to buy opium; but Hawkins knew it was only because of a stringent Government order that they must pretend to obey. He had only as a rule to urge the comity of nations and the old commercial treaties between England and Spain for the Spaniards to buy his dearly coveted wares. If these arguments failed he had another, which at La Hacha was sure to be wanted. Thither the *Judith* and another ship were sent, and at once were fired on. For five days they blockaded the port, and then Hawkins came round with the rest of the squadron. The field-guns and two hundred musketeers were landed, the defences stormed, and the town cleared of Spaniards. At night they began to steal back to trade in secret, Governor and all, and Hawkins did not leave till he had thus sold two hundred negroes.

So the game continued, till the ships were so loaded with gold and pearls that Hawkins would not risk another action, and sailed away northward to take up the Gulf Stream for the homeward voyage. No sooner, however, had he passed the Yucatan channel than two hurricanes shattered his fleet and drove it deep into the Gulf of Mexico. To proceed was impossible without refitting, and he boldly put into Vera Cruz. San Juan de Ulua, as it was then called, was the port of the city of Mexico itself, and twelve large ships laden with gold and silver lay in the harbour waiting for the rest of the Plate fleet and its convoy. They were hourly expected, and next day they arrived off the port to find it in the possession of Hawkins. The whole year's produce of the Indies was thus at his mercy. The galleons within the port were defenceless, and the fleet outside must be utterly destroyed by the first gale unless he permitted it to enter. Never had such a draught been held to an Englishman's lips. But John Hawkins was honest and discreet enough to resist the temptation, and a formal convention was made by which the Spaniards were to be allowed to come in and the English to refit.

Hawkins scrupulously observed the terms of the agreement, but Don Martin Enriquez, the new Viceroy of Mexico, who was with the fleet, had come out with special orders about that "enemy of God," John Hawkins, and he saw too well a road to high favour with Philip. For three days the English were suffered to dismantle their ships, and then, in spite of oaths and hostages and the sacred word of the Viceroy as a gentleman and a soldier, they were treacherously attacked. Though the surprise was complete, a desperate resistance was made. Four Spanish ships were sunk, the flagship reduced to a wreck, over five hundred of their men slain, and at last it was only by fire-ships that Hawkins could be dislodged. The *Jesus*, the *Minion*, and the little *Judith* were all that got clear, and Drake himself, it is said, only escaped by swarming on board along a hawser. The rest of the ships were lost, and so shattered was the *Jesus* that she had to be abandoned with all the immense proceeds of the voyage. Crowded with the crews of the lost ships, riddled with shot, and only half-victualled, the *Minion* and the *Judith* began to stagger homewards, while the Spaniards enjoyed their ill-gotten success. In all those wars it was by far the richest victory which the Spaniards gained over

The King's Library

the English, and of all the most dearly purchased; for not only did it win for Philip and his perjured Viceroy the mortal enmity of John Hawkins and Francis Drake, but it showed them the path to their revenge.

CHAPTER II

THE SPANISH MAIN

IN England the news of the disaster produced a profound sensation. It may fairly be said to mark the opening of a new book in the great epic of the Reformation. For the first time the long commercial intimacy between England and Spain received a rude shock, and from that shock it pined and died. Hitherto the party in the Council that believed England's true policy to be a policy of alliance with Spain had more than held their own; but on January 23rd, 1569, a weather-beaten man was riding post from Plymouth along the London road with the tidings which were destined at last to turn the scale.

That man was Francis Drake. He had been the first to arrive from the perilous voyage. Since the fatal night he had seen nothing of his kinsman. He had put into Plymouth in great extremity, and in spite of his long privations, had been despatched by William Hawkins post-haste to the Council on the spot. It was in a critical moment that he came. Alarmed by the restlessness of the Northern Catholics and the suspicious preparations which Philip's Viceroy, the Duke of Alva, was making in the Netherlands, Elizabeth was "taking care" of a large treasure which had been chased into her ports by the Protestant rovers in the Channel. It was money borrowed to pay Alva's army, and the Spanish Ambassador was loudly protesting. Determined not to let it go, the Queen was yet at her wits' end for an excuse for keeping her hold, and Drake reached the Council doors in the nick of time. At the moment when the Ambassador was pressing his claim with a cogency that was not to be resisted, he suddenly found himself recoiling before a new argument. For the Queen was parrying his home-thrust with demands for an explanation of the outrage offered to her trusty merchant; and, till satisfaction was given, she was flatly refusing to loose her hold on the treasure.

A few days later Hawkins arrived in a worse case than Drake. At first he accused his kinsman of having deserted him, but in the official inquiry, which was immediately held, the accusation was not repeated. He had probably been satisfied with the explanation Drake would naturally have given him, that, as he was already overloaded and short of provisions, he thought it his duty to get home as quickly as possible.

Drake was not present at the inquiry. While he was waiting for the result of the Queen's demands for redress on behalf of her fellow-adventurers, he took service in the navy. The Queen's retention of the treasure had been followed by embargoes on both sides. Trade with Spain was stopped, war with France was imminent, and Sir William Wynter was sent out with a strong fleet under orders to relieve the French rebels in Rochelle and convoy the English merchant fleets to the Baltic, where the swelling trade of the country had pushed a new outlet. It was under him that Drake probably served, and in his school learnt all that the royal service could teach.

In the summer he got leave, and on July 4th was married to Mary Newman, a Devonshire girl, who lived at St. Budeaux close by the town of Plymouth. But he can have had but little leisure to enjoy his new happiness. For England was passing through one of the greatest crises of all her history, and every one knew it. Men believed that the ensuing year would decide the fate of the Reformation, and a presentiment of coming evil hung over the nation. In the winter the insurrection of the Northern Catholics prematurely exploded, and though it was easily smothered, the success of the Government did little to relieve the situation. The dawn of the year 1570 was darkened with the threatening shadow of a crusade for the release of Mary Stuart. England resounded with warlike preparations; the forces of the south and west were mobilised; every available ship was being brought forward for service; and John Hawkins was equipping a force which was supposed to be under orders to intercept the Plate fleet at the Azores. We see all England at this moment agitated by the rough gusts that herald the storm; we see her overshadowed with war-clouds; the horizon blackens with coming danger, and like a gleam of hope the white sails of two small vessels scudding westward out of Plymouth harbour detach themselves brightly against the surrounding gloom.

The secrecy which shrouded this daring expedition is still impenetrable. It was only known that Francis Drake had sailed with two small vessels, the *Dragon* and the *Swan*, to reconnoitre the Spanish Indies till he found at the very well-spring of her life a point where some terrible wound could be inflicted on England's enemy. In England it was always believed to be a private venture for revenge, which began and ended with Drake himself. Like Hawkins, he was burning to rescue from the clutch of the Inquisition the comrades of the last voyage who had fallen into the hands of the Spaniards. Diplomacy had failed to obtain redress for their losses, and both were bent on reprisals. Still, when we consider the demand there was at this moment for naval officers, with what care the intelligence department was being organised, and finally how the expedition was prepared under the eyes of Drake's old patron, William Hawkins, who was now Governor of Plymouth, it is impossible to believe that the Government was not in some way concerned. Without the incentive of a special mission Drake would hardly have sailed while his old commander was hourly expecting to be loosed on the Plate fleet; and the truth most probably is that Hawkins had suggested the possibility of reprisals in the Indies, that his idea had been favourably received as being less likely to lead to an open rupture than action in Europe, and that he had employed Drake to secure the necessary information.

In the following year the expedition was repeated with the *Swan* only. Whatever Drake did and whatever he saw, the effect of these two adventurous voyages was to earn him a reputation for humanity with the Spaniards and beget in him a magnificent contempt for their power. He returned with his head full of a scheme so wild in its daring that in the bare contemplation of its extravagance we seem transported into the world of the Seven

Champions. Spain bestrid the world like a Colossus, half Europe was crawling between its legs, and Drake was volunteering with a jest on his lips to steal the hen that laid the giant's golden eggs. The two years of danger through which the State had just passed had bred a spirit ripe to applaud such an enterprise. Elizabeth's excommunication, and the discovery of the great Ridolphi plot, had filled her with an ugly desire for retaliation both against Rome and Spain, and the relations between London and Madrid were strained once more to breaking-point. Although at last she wisely smiled upon Alva's suggestion that they should kiss and be friends yet a little while, an adder's poison was under her lips. In deference to the Proconsul's wishes she refused any longer to harbour the savage De la Marck and his Beggars of the Sea under the guns of her Cinque-Ports; but the Governor of Plymouth received no orders to stay two wicked little craft which under his very eyes were being fitted out with every device which the latest science could suggest.

For Francis Drake had found friends to back his wild scheme, and on Whitsunday eve, May 24th, 1572, he sailed out of Plymouth Sound on board the *Pasha*, of seventy tons, and in his wake was the little *Swan*, of twenty-five, in command of his brother John. Another brother, Joseph, was with him too, and John Oxenham of tragic memory, and others whose names were destined to be not entirely lost in the coming blaze of brilliant reputations. The Spaniards always said the Queen was among those who subscribed the cost, but be that as it may, no ship in her navy was better furnished than these. In all respects they might have been Her

Majesty's own men-of-war, and yet the whole project wears the air of a schoolboy's escapade. The crews all told, men and boys, numbered but seventy-three souls; there was but one of them had reached the age of thirty, and their modest end was nothing less than to seize the port of Nombre de Dios, and empty into their holds the Treasure-House of the World.

On his previous voyage in the lonely depths of the Gulf of Darien, Drake had discovered a little landlocked bay.[1] Here he had buried his surplus stores, and here on July 12th he hove-to his tiny squadron. Secure in the solitude of these untrodden shores, he intended at his leisure to set up three dainty pinnaces which he had brought from Plymouth in pieces, and with which the attack was to be made. No sooner, however, had his boat passed the narrow entrance of his hiding-place, than he saw smoke rising out of the dense tropical forest in which it was embosomed. His fastness had been profaned, and after returning to the ships for more strength and arms, he went ashore and boldly plunged into the tangled vegetation to solve the mystery. Not a soul was to be seen, but a leaden tablet met his view, informing him that some prisoners whom last year, in defiance of all precedent, he had released instead of drowning, had betrayed his hiding-place, and that the Spaniards had removed his stores. One of his former crew had conducted a rover called Garrett to the spot, but, in alarm to find the place had been discovered by the Spaniards, he had hurried away, and the only traces

[1] Probably the lower Puerto Escondido (Hidden Harbour) marked on Keith Johnston's map.

left of him were the kindly warning on the leaden tablet and the smouldering fire.

But Drake was not so easily daunted. He had fixed on the spot for setting up his pinnaces, and he did not mean to leave it till they were afloat. The ships were brought in, a pentagonal entrenchment marked out on the shore, and all hands were soon hard at work clearing the ground about it and hauling the felled trees together to make an abattis for its defence.

So protected, the carpenters next day got to work on the pinnaces, but it was only to be interrupted by the appearance of a suspicious squadron bearing straight for the harbour. It consisted of a ship, a caravel, and a launch or shallop. It proved to be nothing worse than a vessel of Sir Edward Horsey's, Captain James Ranse, with two Spanish prizes in company. Ned Horsey had been a notable pirate in the old wild days when Mary was on the throne, and any man in his employ was not likely to prove bad company for a venture like the one in hand. No sooner had Ranse heard what was in the wind than he begged to stand in with Drake, and the two rovers, with all the solemnity of ink and wax, drew up a deed of partnership.

In a week the pinnaces were ready, and the combined squadron stole out of its hiding-place. Proceeding north-westward along the coast, on the third day they reached some fir-clad islands, which they called Islas de Pinos. Here, too, they found themselves forestalled. Some negroes were in possession, loading planks and timber into two frigates or small galleys from Nombre de Dios. The men were seized and eagerly questioned for news of the state of the town. Their information was as bad as it could be. The waste of hill and forest that lay on either side of the road from Panama to Nombre de Dios was occupied by a savage black people whom the Spaniards called *Cimarrones*, a word our seamen corrupted into Maroons. Eighty years ago a number of African slaves had been driven by the cruelty of their masters to take to the woods, and having found favour in the eyes of the Indian women, they had now grown into two great tribes, whose terrible mission it was to rob, and kill, and torture every Spaniard on whom they could lay their hands. Filled with the savage cruelty natural to so mongrel a race, possessed of a splendid physique, and alarmingly prolific, they were dreaded and treated by the Spaniards like wild beasts. So formidable had these terrible tribes become, that this very year they had annihilated a strong force which a chivalrous Spanish gentleman had volunteered to lead against their stronghold. Six weeks ago they had almost succeeded in surprising Nombre de Dios itself, and the Governor in great alarm had sent to Panama for reinforcements.

Nothing could have fallen out worse. The town on whose sleepy security the success of the expedition depended would be all on the alert, and without shame Drake might well have reconsidered his determination. But for him the news was but a reason for immediate action, on the chance that the troops had not yet arrived from Panama. With that remarkable foresight which seemed always to temper his rashest moods, he set the negroes ashore, trusting they would find their way to the Maroons with a good report of his kindness, and, at the same time, made rapid preparations to reach

Nombre de Dios before they could report his presence on the coast, in case they were so minded. It was arranged that the three ships and the caravel were to lie hidden amongst the Pine Islands under Ranse, while Drake himself made the attempt with the three pinnaces and the shallop. Fifty-three of his own men and twenty of Ranse's were picked for the service, and on the 23rd the last farewells were said as the four boats rowed off on their desperate adventure.

The arms as yet were lying snugly packed in chests. In their selection Drake showed a scholarly respect for the latest ideas of infantry officers, no less than that almost humorous originality which is so characteristic of all his exploits. Pikes and firearms were in equal proportion, according to the approved practice of the time, and the officers were armed with sword and buckler; but with the remaining third of his force he permitted himself the indulgence of an ingenious fancy. Sixteen men were armed with bows, and supplied not with heavy war-arrows but with light roving shafts, specially devised to produce the same distracting effects for which rockets were afterwards employed. Six of the pikes, too, were fitted with gear for holding blazing tow, while the remaining four men carried nothing more deadly than trumpets and drums for the sole purpose of making as much noise as possible, as well for the encouragement of their comrades as the alarm of the enemy.

In five days they had covered twenty-five leagues and reached a group of islets which was known to them as Cativaas or Prisoners' Islands. Here at break of day the men were landed, and paraded in their respective parties, while Drake with cheery speeches served out

their arms and did his best to remove the bad impression which the negroes' news had made. In the afternoon they were away again, and by midnight the four evil-looking craft were lying still as sharks under the point Nombre Bay. Hither, creeping stealthily along the shore, Drake had conducted them unperceived, and here they must wait for the first gray streaks of dawn. It was a time to try the stoutest heart. As the dark hours dragged wearily on, the young hands grew more and more nervous. On the other side of the point under which they lay was a world-renowned town, as big as Plymouth, by this time perhaps full of the unconquered Spanish infantry. There was nothing to break the spell of the deathlike silence but the booming of the surf and their own anxious whispers, as they discussed beneath their breath the negroes' news. Drake knew that another hour of such suspense would wither the heart out of his force, and it wanted yet an hour to dawn. Suddenly he descried a faint light silvering the horizon. It was only the moon rising, as he well knew, but by a happy inspiration he proclaimed it the dawn, and cheerily ordered out the sweeps.

No sooner was the harbour gained than they knew they were detected. As ill-luck would have it, a ship was just taking up her moorings, and to their dismay a boat shot from her side straight for the landing-place. In a moment Drake's pinnace was leaping across the water at racing speed to head it off. It was a desperate struggle, but Devon muscle told. The Spaniards soon saw they were overmatched, and fled to the opposite side of the bay. A few minutes later the four boats had grounded under the shore-battery, and the men were tumbling the guns into the sand. Unfortunately the gunner in charge had escaped to give the alarm, and as the men hastily fell into their places, they heard the citizens take fright. No sound under heaven is more awe-inspiring than that of a town awakening in terror. The first confused murmur that is quickly broken with scattered cries, the first woman's shriek and first excited shout, each again and again repeated till the shuddering murmur is a broken roar—all this beset the ears of these threescore sailor lads, and worse than that, the roll of drums calling troops to arms, and the din of the great church bell high above all.

But now there was no time to be afraid. Twelve men were left to guard the boats, and the rest hurried silently to seize the new eastern battery. It was Drake's great anxiety, and to his intense relief he found that not a gun had yet been mounted. So now the real work could begin, and divided into two parties, with clatter of drums and blare of trumpets, brandishing their blazing fire-pikes and uttering horrible yells, they rushed by two different ways to the market-place. Drake with the bulk of the force ran up the main street, while his brother and Oxenham led the smaller party by a more devious route with which they were somehow familiar. At the corner of the Plaza where the Panama road left it a considerable force had by this time managed to assemble, and as the head of Drake's little column appeared it was received with a roar of musketry. The sand spat up about their feet: the trumpeter dropped; and the rest with one volley of shot and arrows dashed at the glowing matches before them. Then hand to hand, with sword and pike and the swinging butts of muskets, the doubtful struggle raged. The event yet hung on the balance, when in upon the Spaniards' flank burst the second party with a startling volley. Bewildered with the darkness and the din the Spaniards' hearts began to fail. Panic multiplied the handful of their foes. Maddened by the roving arrows, scorched and blinded by the flaming pikes, they were hustled like sheep into the narrow road. Faster and faster they gave their ground, arms were flung down and backs were turned, till all in hopeless rout were flying for life through the Panama gate.

The Plaza was immediately secured at all its outlets, and Drake with a small guard made a move for the Governor's house. No sooner had they entered the storehouse than a sight of wonder met their eyes. Against the wall they saw the gray shimmer of a pile of silver bars ten feet in breadth, twelve feet in height, and seventy in length. It was a sight beyond the most fevered of the Devon lads' dreams. In open-mouthed amazement they prepared to fall upon it, but Drake only laughed at them and gave a sharp order to stand to their arms. He knew the danger was yet extreme. The town was full of soldiers, and to remove the silver in face of them was madness. He sternly forbade a single bar to be touched, for he knew that down by the shore was the King's treasure-house, where was a store of gold and jewels beside which the heap of silver was as mere ballast.

Returning to the main body he found that the reaction from the excitement of the fight had brought the men back into their nervous state. The distracting clang of the great alarm-bell was still crying lustily for help, for Drake would not permit the church to be profaned; inky clouds were rolling up over the town and obscuring the moon; frequent shots and warlike cries had been heard by the shore, and the rumour was that the boats had been attacked. John Drake and Oxenham were at once ordered off with their party to ascertain the truth and then rendezvous at the King's treasure-house. Thither Drake hurried with the rest, but ere the goal was reached the tempest burst, and to all their dangers and terrors was added the unearthly roar of a tropic rain. Matches were quenched and bowstrings ruined before they found shelter in a shed attached to the treasure-house. Here another insupportable delay occurred, for from the spot where they stood it was impossible to break in. In the midst of the growing dismay John Drake and Oxenham ran up to report the boat-guard safe but in great alarm at the numbers of soldiers they had seen hurrying to and fro. A friendly negro whom they had taken on board reported that a hundred and fifty soldiers had arrived the day before, and that unless the English got clear before dawn they would certainly be overwhelmed. In the face of this it was impossible to keep the young hands steady many minutes longer. In vain the undaunted captain tried to occupy them in repairing the damage which the rain had done to their arms; in vain he encouraged them with hopes and even taunts. "I have brought you," he cried, "to the mouth of the Treasure-House of the World. Blame nobody but yourselves if you go away empty." The men only grew more unsteady; and as the rain abated a little Drake ordered his brother and Oxenham to go round and break open the treasure-house door while he returned to hold the Plaza. Briskly he stepped forward to lead the way, and then with a cheery cry on his lips rolled over in the sand.

Ever since the first encounter he had been struggling against a desperate wound in the leg, lest the timid should make it an excuse for returning with the plunder they had already collected. Now they saw that his very footprints were full of blood, and it was clear his life was in jeopardy. The boldest would not listen to his entreaties to proceed with the work. His life, they said, was dearer to them than all the wealth of the Indies. In spite of his angry protests they bore him with loving violence on board his pinnace, and so as the four boats rowed moodily from the harbour the sun rose on their glorious failure, and the renowned attempt upon the Treasure-House of the World was at an end.

Every colonial port in those days had its Isla de Bastimentos or Victualling Island, and to that of Nombre de Dios the boats made their course, taking with them for their comfort the ship which they had seen arrive. It was laden with good Canary and other stores, and with the aid of this and the abundance of delicacies on the island Drake and the rest of the wounded rapidly recovered. Hither they were followed by one of the Panama officers, who professed himself overwhelmed with the brilliancy of Drake's feat of arms. He wanted to know if he was indeed that incomprehensible rover who did not drown his prisoners; and if so, if there was anything with which the Governor could provide him. Drake, who took the Don's fantastic courtesies for a cloak to cover a spy, answered roughly enough that he was the Drake they meant, and that he intended to help himself to all he wanted. "So tell your Governor," he concluded, "to hold his eyes open. For before I depart, if God lend me life and leave, I mean to reap some of your harvest which you get out of the earth and send into Spain to trouble all the earth."

Drake meant what he said. He was still unsatisfied; he was more contemptuous of the Spanish power than ever, and his men were intoxicated with their leader's spirit. But not so his partner. When the combined force was once more assembled at the Pine Islands, Ranse declared he was not willing to risk the booty already obtained in the hornets' nest they had stirred. So with mutual goodwill they wound up the partnership and parted company. Ten days of rest had elapsed since the attack on Nombre de Dios. Their wounds were already half forgotten, and as Ranse shaped his homeward course, Drake swooped upon Cartagena, the capital of the Spanish Main. As he approached, the boom of guns rolled out across the sea to tell him he was expected. Light airs had delayed him, and in spite of his promptitude a despatch-boat had slipped in to give the alarm. Nevertheless, with his pinnaces he boldly entered the harbour, cut out a large ship that was discharging at the quays, and amidst a loud alarm of guns and bells and musketry carried it in triumph out to sea. Next day he intercepted two more despatch-boats, and learned the full extent of the tempest he had raised. Dazzled though he was with his recent feat, he saw clearly the ferment must be allowed to subside. With bewildering suddenness his whole plan was changed, and next night no trace of him was to be seen off the

Spanish Main but the charred remnants of the *Swan* burning down to the water's edge.

He had utterly disappeared as suddenly as he came, and the mystery of the burning ship was all he had left behind. It was a mystery the Spaniards were not likely to fathom. For the new project on which his genius was bent the pinnaces had to be fully manned, and to man the pinnaces his brother's ship must be abandoned. His extraordinary instinct for the control of men told him that to give such an order was but to court mutiny; his clear-eyed tenacity of purpose told him the work must be done, and done cheerfully. At such moments his influence over men was miraculous. That night Tom Moone, the carpenter of the *Swan*, was secretly scuttling his beloved vessel at the risk of his life, and in the morning she was half full of water; but so skilfully had Drake ordered the work that no leak could be found. All day he toiled with well-feigned anxiety at his brother's pumps till he and all the company were exhausted, and the water hardly reduced. In despair her heart-broken young captain sought his brother's advice, and so astutely was it given, that that night, as the shadows fell, John Drake with his own hand fired his stricken ship.

CHAPTER III

THE MULE-TRAINS

To the Spaniards, that sight could only be a sign that the redoubtable rover had left the coast. The truth was far otherwise. On the western shore of the Gulf of Darien a little natural harbour, which through the ages had been undisturbed, was suddenly teeming with busy life. Its primeval silence was awakened with the jolly laugh of the bellows, the ring of the anvil, and the snap of the axe, and the monkeys watched with worried air the mysteries of bowls and quoits and archery, and all the pastimes of an English May-day fair. A wide stretch of forest growth had given place to rows of leafy cottages, and by the shore a little dockyard waked the echoes with unceasing clatter. It was another Puerto Escondido which Drake had discovered, and here, far from the ken of Spaniards, with alternate days of work and play, he was refreshing his men and refitting his squadron.

They called the place Port Plenty, for from hence they swept the sea of every passing victualler, and added the captured cargoes to the stores of game and fish it was their delight to catch. At intervals along the coast and amongst the wilderness of islands magazines were hidden, and into these were poured the stores that had been destined for great Plate fleets. The shark-like pinnaces would suddenly appear in the midst of the trade-route no one knew whence, and, laden with food, as suddenly disappear no one knew whither. Even the solitudes of the far Magdalena river beyond Cartagena were startled with the splash of English oars; and ere resistance could be made, the storehouses on its banks were swept clean, and all was silence once more.

It was on his return journey from this expedition that Drake learnt the first step towards the new exploit was accomplished. With the help of Diego, the negro who had joined at Nombre de Dios, John Drake had succeeded in establishing communication with the Maroons. While the Spaniards were straining their eyes seawards, Drake was quietly preparing to seize King Philip's gold behind their backs as it left the gates of Panama. To this wild project the assistance of the Maroons was essential, and the first negotiations convinced him how eager they were to help. But they had to assure him that nothing could be done till the dry season, for during the rains no gold was conveyed by land. Weary months must elapse before the blow could be struck—months of inaction, which in that terrible climate Drake knew was his greatest enemy. Everything was done to mitigate the evil. The Maroons showed them a new and more convenient harbour than their last, somewhere, it would seem, amongst the shoremost isles of the Muletas. Here, within a network of reefs, the vessels were snugly ensconced, and with the eager help of the Maroons a great fort of timber built.

In October, leaving his brother John as governor of his little kingdom, Drake with two of the pinnaces sailed once more for the Main in search of intelligence. For more than a month every kind of bravado was indulged in, partly, it may be, from a boyish delight in putting indignities on his enemy, and partly from a sagacious purpose of keeping up the hearts of his men. For in Francis Drake, as in some hero of saga, reckless high spirits and a far-sighted wariness seem ever contending for mastery. He gathered fruit in the gardens of Santiago de Tolu; he cut out vessels from the very mouth of the Spanish guns; he rode out two gales in the harbour of Cartagena itself; he jumped ashore in the face of the garrison to show how cheaply he held the arms of Spain; and yet with a skill and judgment beyond his years he eluded every attempt to capture him by force or fraud; and through sickness and hunger, through exposure and disappointment, he maintained his men in such a state of cheerful obedience as had perhaps never before been seen.

And now, as though to teach the world what heroic fortitude, what a depth of patient resistance, lay beneath the tumultuous surface, misfortune came upon him apace. He returned to find his brother dead. In spite of his orders, Captain John Drake had been taunted by one a little more insane than the rest to attack with no better arms than a rapier and an old firelock a frigate full of Spanish musketeers. It was certain death, and both of them fell at the first discharge. The whole company were overwhelmed with grief for the loss of a man whose brave and loyal nature seemed destined to fill a place but little lower than his brother's. It was their first reverse, and it came at an untimely moment. It was now December. The rains were ceasing; next month the Plate fleet would arrive from Spain; the gold would begin to move; and the time for their daring land-journey was drawing near. It was necessary that they should keep perfectly quiet till the trusty Maroons brought them news that the fleet had come. In painful inaction, therefore, they had to bide their time, while each day the heat increased as the dry season set in, and pestilence steamed from the sodden soil. December dragged wearily to its end, and with the dying year the vitality of the men ebbed fast away in the poisonous atmosphere. With the first days of the year 1573 ten men dropped in a raging fever, and in a few hours half the company were down. Death succeeded death, and the surgeons were helpless. Joseph Drake was seized amongst the rest, and expired in his brother's arms. Then Drake arose as valiant against the ghostly enemy that was mastering him as he ever showed himself in open fight. Burdened as he was with the horrors of that fever-stricken den, and revolting and inhuman as the bare idea of dissection then was, he resolved to violate the sanctity of his own brother's corpse to save the rotting remnants of his company. The weight of prejudice and tender sentiment he had to fling aside, to us is hardly conceivable, but the memory of it seems to send a shudder through the chronicler as years after he told the tale. "It was the first and last experiment," he exclaimed, "that our Captain made of anatomy ir this voyage."

It was the end of January before the Maroons' scouts reported that the fleet had put into Nombre de Dios. A pinnace was at once despatched to test their fidelity, and it returned with a victual-frigate, whose captive crew and passengers confirmed the negroes' intelligence.

All was now bustle and excitement. Of the seventy-three souls who left Plymouth eight months ago but four-and-forty now remained. Half of these were too fever-stricken to march, and some sound men it was necessary to leave behind to tend the sick and guard the prisoners from the fury of the Maroons. Eighteen were all that could be spared for the adventurous journey, and on Shrove Tuesday, February 3rd, they plunged into the forest with thirty Maroons in company. On the third day they reached a stronghold of their allies, where they were overwhelmed with hospitalities and offers of assistance. But Drake would not tarry or accept any increase in his force. Yet his heart was none the less moved with their simple kindness, and with pity when he saw them making a fetich of the cross. His earnest Protestantism would not suffer him to leave them in such a slough of sin, nor did he depart till he had persuaded them to cast away their fetich, and in its stead had taught them like children to say the Lord's Prayer, and fed them with some scraps of the old navy chaplain's divinity. Then with his heart lifted to heaven he strode on once more to spoil God's enemies.

Their march had now begun in earnest. Not a word was spoken; with all the breathless caution of the chase they followed their strange familiars up the forest-clad spurs of the Cordilleras. A mile ahead four guides felt their way, as it seemed by magic, through the sunless forest, and marked the track with broken boughs. Twelve more Maroons headed the little column, and twelve more formed its rear-guard. In the midst with the two black chiefs marched the Englishmen. To them it was like some enchanted land. Their devoted friends would not suffer them to carry the lightest burden, and at their ease they crept along the trail in silent wonder. The miraculous instinct of their guides, the beast-like skill with which the hunters killed their game, the fascination of the endless silence, the wealth of luscious fruit, the prodigies of vegetation, the wonderland of birds, all mingling confusedly with the half-forgotten thought of the wild adventure before them—it was enough to make their lives seem turned to fairy tales, without the vision that was at hand. On the fourth day after leaving the stronghold, the spot on which every thought was bent had been attained. It was at the summit of the range. High upwards rose a giant of the forest, and in its arms was a leafy bower. Drake first ascended, and sank awe struck upon his knees. First of all Englishmen he was gazing on the Golden Sea. Before him spread the unmeasured mystery of the Pacific; at his back rolled the old Atlantic. His heart was overflowing—so Moses gazed upon the promised land—and like a good crusader he humbly besought Almighty God of His goodness to give him life and leave to sail once in an English ship in that sea. So he descended and told them of his prayer; and before them all John Oxenham, his lieutenant, vowed that unless the captain did beat him from his company, he would follow him, by God's grace.

The secret of that fabled ocean, whose very name for two centuries thereafter never failed to inflame the imaginations of high and low with dreams beyond the extravagance of alchemy, was a secret no more. Jealously as Spain had veiled the South Sea's beauty,

Drake had seen her face, and athirst with desire he began the perilous descent to Panama. In two days the shelter of the forest ceased. The open country increased their danger every hour. On Valentine's Day the magnificent roadstead of Panama with its burden of gold-ships opened before them, and now the peril of discovery became acute. Fearing at every step to be surprised by some fowler from the city, they broke into twos and threes, and so by different routes crept like lynxes through the giant grass, till a grove a league outside the gates gave them a semblance of security. Here, beside the Nombre de Dios road, they lay concealed, and rested while a spy was sent in disguise into the city to get news of the movements of the gold-trains. Drake had hardly finished from the skirts of the grove a hazardous reconnaissance of the city against the time when God should grant his prayer, than the spy returned bursting with news. Two large mule-trains laden with victuals and silver were getting ready to start in the market-place, and in front of them that night was to travel no less a man than the Treasurer of Lima himself on his way to Spain, with eight mule-loads of gold and one of jewels.

The luck indeed seemed turned at last, and in high hope, as night was closing in, a move was made for Venta Cruz. Here, at the point where the Nombre de Dios road crossed the river Chagres, stood the half-way depot, and thus far the mule-trains always journeyed by night, so fierce was the noonday heat across the grass-clad prairie. Thence, either by the river or the roughly paved road which Pizarro had made through the cool forest, the treasure reached the Atlantic. Within a couple of leagues of the town they halted and prepared their ambush. First every man pulled his shirt over his clothes as the art military demanded for night attacks; and then, divided into two equal parties, they took up their places on either side of the way, some fifty yards apart. By this disposition the first and hindmost mules of the train could be seized at the same moment, and each party could use its weapons against the Treasurer's escort without hurting the other. Thus with every detail arranged, and certain of their prey, they waited.

For an hour the silence of death lay upon the grass-bound highway, broken only by the uneasy breathing of the crouching ambuscade, and the murmurous voices of the tropic night. Then faintly over the soft rustle of the giant grass came the tinkle of mule-bells on either hand, and from the direction of Venta Cruz the sound of a horseman's trot and a running foot-boy. The strictest order had been given that everything going towards Panama should be let by without a move, and unsuspecting the traveller came on at an easy pace. Suddenly amidst the waxing sound of the mule-bells Drake to his dismay heard the trot change to a canter, and the gentleman clattered rapidly down the hard road towards Panama. Still, it was impossible that he could have seen anything through the impenetrable grass, and the anxious captain lay quietly on. Surely enough the mule-trains were not alarmed. Louder and louder jangled the deep-toned bells till the air was alive with the merry clang. Then sharp and high over all Drake's whistle pierced the din—the grass bowed before the rush of black and white figures—oaths and curses mingled with the bells—the mules stopped and then lay down—and almost without a blow the whole train was in Drake's hands. Pack after pack was rent and ransacked of its contents. A little silver was found, but Treasurer there was none, nor a single jewel, nor an ounce of gold. It was from the chief muleteer they learned the bitter truth. A sailor who had been keeping up his courage with *aqua vitæ*, had been fired to spend it on the traveller. His Maroon mate in a moment had knocked him down and lain on the top of him, but in the struggle they had rolled into the road. The gentleman had galloped on in great alarm, and meeting the Treasurer had told him that El Draque—how and whence he could not tell—was lying in wait for him on the lonely road, and the Treasurer had consented to send on the victual-train in front of him to spring the trap. Thus early did that ominous name begin to have its unearthly sound in Spanish ears. Time and space seemed already not to exist for him, but in truth they were now a terrible reality. Delighted as he always was to hear how he frightened superstitious Spaniards, it was no moment for any but the gravest thoughts. Horse and foot would soon be hurrying out of Panama, and in Venta Cruz perhaps the alarm had already been given. The whole party were much exhausted by their recent long marches. To retrace their steps to a place of safety was four good leagues, and the only other line of escape lay through Venta Cruz.

Needless to say the latter course was chosen. Each man mounted on a mule, they continued their way, till close to the town the Maroons scented musket-matches amongst the trees ahead. Dismounting, they boldly held on along the narrow road till they were challenged by a Spanish officer. In the name of the King of Spain the pompous captain bade Drake yield. For the honour of the Queen of England, the seaman cried, he must have passage that way, and flashed his pistol in the Spaniard's face. It was the signal for a volley from the Spanish musketeers. As the firing ceased Drake's whistle rose merrily once more, and then through the choking smoke sailor and Maroon dashed blindly at the unseen enemy. Yelling and leaping like fiends the negroes lead the charge. *Yó pehó! Yó pehó!*—their terrible war-cry struck dread into the Spaniards' hearts. Backed by the maddened mariners the charge was irresistible. Without a check the enemy were swept through the town-gate up the narrow street and into the monastery—monks, soldiers, and civilians in a panic-stricken flock—and there they were safely locked while the victors pillaged the little town.

Besides its commercial importance, it was a sanatorium to which the ladies of Nombre de Dios came to be confined. Drake had reissued his invariable orders that no woman or unarmed man should be touched, and even in the heat of the sack his savage allies did not attempt to disobey. Yet the poor distracted invalids never ceased their piteous entreaties until Drake himself came to their bedsides to comfort them. Surely never was a pirate so tender, or with such a gentle name.

For an hour and a half the pillage continued, till an alarm of horse interrupted it. It was the advance guard of the cavalry from Panama; but a picket was holding the gate, and so well had Drake kept his men in hand that, faster than they came, the new-comers were soon galloping back to join their comrades. When they returned with the main body to storm the captured town the corsair had vanished.

Far and wide the alarm spread. To the confines of Nicaragua the mine-owners did not feel safe, and made ready for a flight to the South Sea. Within a fortnight one of the shark-like pinnaces appeared in the port of Veragua, and it was only the unsleeping vigilance of the garrison that saved a vessel laden with a million of gold from Drake's hands. A frigate of Nicaragua put in with news that Drake had boarded her and stripped her of her gold and her Genoese pilot. To the eastward his lieutenant had captured and carried off a fine frigate laden with victuals. No one, in short, knew where to look. All that was clear was that he was at sea again, and the gold-frigates hardly dared move.

Meanwhile the treasure was pouring across the Isthmus under strong escorts unmolested. On the last night of March, guarded by half a company of soldiers, three large mule-trains left Venta Cruz with some thirty tons of silver and a quantity of gold. Almost to the very gates of Nombre de Dios they travelled as safely as the rest, when just as they thought all danger over, in front and rear the jangle of the bells was drowned in a rattle of musketry. The woods were belching shot and arrows, the air was rent with that terrible *Yó pehó! Yó pehó!* mingled with shouts in French and English. Overwhelmed and dazed as the yelling figures leaped down upon them into the road, the soldiers discharged their pieces and fled into the city in wild alarm. The garrison turned out with all the promptitude which their constant watchfulness made possible; but when they reached the scene there was nothing but the mules and the empty packs.

It was Drake again where least of all he was to be expected. Shortly after his attempt to cut out the gold-frigate of Veragua he had met with a Huguenot privateer in distress. He had relieved its wants, and had heard from its captain the news of the Massacre of St. Bartholomew. Moved with pity at his heartrending tale, and lashed into a fury against the whole Catholic world, Drake had taken him into partnership and determined on this last desperate attempt. Strong in the certainty of his simple faith that God's hand must surely now be with him against the idolatry that was stained with a crime so hideous, he had struck his triumphant blow at the gates of this Moloch, and what wonder if he thought it was the finger of the Lord that had pointed out the way?

While the Spaniards were gazing hopelessly at the rifled mules the victors with jests and laughter were stuffing silver bars into the land-crabs' holes, hiding them under trees, burying them in the river-bed, till fifteen tons were concealed. Then groaning under the rest and all the gold, they staggered on to the river-mouth, where the pinnaces were to meet them. Overtaken by a storm of rain, in two days they reached the rendezvous drenched and exhausted,—and there, with the suddenness of a dream, at the very culmination of their fortunes, they found themselves face to face with a situation beside which all their former griefs were as nothing. Not a sign of the boats was to be seen, and in their stead appeared seven Spanish shallops. They were rowing towards Nombre de Dios from the very spot where the pinnaces had been lying during the absence of the shore-party, and all hope, not only of saving the treasure, but of ever seeing home again, was gone. The pinnaces must have been overpowered, and under torture the prisoners would have to confess where the ships were hid. Despair seized every heart but Drake's, and invoking the aid of that extraordinary power in which he was never surpassed, with a few cheery words he transformed the situation into one of hope. He showed them that if God had permitted the enemy to prevail against the pinnaces, He had sent the storm to bring tree-trunks down the river; with these they might make a raft and reach the ships long before the blundering indolent Spaniards could make up their minds what to do. "It is no time to fear," he urged, "but rather to haste to prevent that which is feared." In a trice all was movement again. The raft was made; and with a crew of one Englishman and two Frenchmen, who insisted on sharing the danger, with a biscuit-bag for a sail and a tree for a rudder, Drake was waving a cheery adieu to his company. "If it please God," he cried, as they pushed off into the stream, "that I shall ever set foot aboard my frigate in safety, I will, God willing, by one means or another get you all aboard in despite of all the Spaniards in the Indies."

Yet no one knew better than he how desperate was his case. No sooner were they at sea than every wave surged over the crazy raft up to their arm-pits as they sat. For six hours they toiled, parched and blistered with the salt and sun. "See," suddenly cried Drake in the midst of their torment, "there are our pinnaces." True enough they were there, and to the sufferers' intense joy bearing straight for them. Drake declared all fear was over; when all too soon it was clear the pinnaces had not seen them, and in a few moments they disappeared behind a small headland, evidently meaning to stay there the night. Without a moment's hesitation Drake steered his raft straight ashore through the raging surf. Reckless as the resolution was, the landing was safely effected, and running round the headland they quickly joined the boats. The crews were horror-stricken to see their disfigured Captain with so ragged a following, nor would he relieve their anxiety till with cruel jocularity he had grimly enjoyed their dismay. It was not until he had ascertained that it was the gale which had caused their delay that he had mercy on them. "Give thanks to God," he cried at last, as he drew from his bosom a quoit of gold, "our voyage is made!"

Well he might say it. The shore-party without further adventure was brought off safely with all their treasure, and much of that which had been buried was subsequently recovered, in spite of the Spaniards' efforts to find it. The booty they now had must have been very large. Besides the plunder of Nombre de Dios, Venta Cruz, and the mule-trains, they calculated that of the two hundred vessels of all kinds which then navigated the Caribbean Sea, there was not one they had not overhauled once at least, and some of them had suffered three times. Their only thought now was the homeward voyage. So reduced was the company that the *Pasha* was too large for them to navigate, and Drake gave it to his Spanish prisoners as some compensation for their long detention. In the Magdalena he knew he would find plenty of craft to serve his turn. So after parting company with the French on the best of terms, he sailed in his new frigate for the river. In Cartagena harbour were all the great Plate ships and their convoy on the eve of sailing for Spain. But in the exuberance of his spirits Drake stood close in, and then ran by before the whole fleet with the flag of St. George waving defiance at his mast-head, and his silken pennants and ensigns floating down to the water to bid them a mocking farewell.

Another frigate was soon picked up, and after lingering in his stronghold long enough to refit for the voyage and take leave of his trusty allies, he shaped his course for home. So prosperous was the voyage that he did not even touch for water at Newfoundland, as the custom then was; and on Sunday, August 9th, 1573, the good folk of Plymouth, to their preacher's dismay, came running out of church as the triumphant young captain's guns thundered a salute to his kinsman's batteries.

CHAPTER IV

GLORIANA AND HER KNIGHTS

THE storm-clouds had rolled away, and the sun was shining peacefully over England, when Drake came home. A week ago the last adherent of Mary Stuart had been hanged in Edinburgh: Alva had turned his back upon the discredited English Catholics; and Elizabeth's Government was drawing a long sigh of relief. Both England and Spain were but too glad to enjoy the respite which Alva's overtures afforded, and Drake, burning with his desire for the South Sea, found himself plunged into the chilling waters of an amicable arrangement. So far from there being any hope of the Government countenancing his passion, he was confronted with the possibility of being sacrificed as a pirate on the altar of Peace. Elizabeth was surrounded by friends of Spain; the master of her household was in Philip's pay; Mr. Christopher Hatton, her new favourite, was a Papist in his politics; and although the Ridolphi plot had ruptured diplomatic relations with the Escurial, there was a recognised Spanish agent in London, who began protesting with such energy that Drake found it necessary to hurry out to sea again for fear of arrest.[1]

[1] See draft memorandum of answer given to complaints of Spain,

There is much to be said for the Spaniard's contention that his proceedings were flat piracy. True, there is no definition of the offence accepted by modern publicists which would fix the guilt upon him. He had not acted as a *hostis humani generis*, but as an enemy of Spain alone; he had not pillaged *animo furandi*, but under colour of right. Still at that time International Law had not so nicely ascertained the limits of piracy and irregular reprisal. That Drake was entitled to reprisal under the old Law of Nations, there was no denying. He had been wronged, he had applied through his Government for redress, and redress had not been forthcoming. But it was now a well-established doctrine that before a subject of one country put in force his remedy of special reprisal against the subjects of another, he ought to obtain a commission from his own Prince, or some authorised official of his Government. Such a commission John Hawkins held at this moment, and William Hawkins was the Queen's Governor of Plymouth. If Drake was not employed by the one, he at least had the connivance, if not the sanction, of the other. Whether Elizabeth had directly authorised the expedition, the friends of Spain could never find out. They blew hot and cold on the point, like men with a weak case. At one moment they tried to fix her with complicity, in the next they sought to convict Drake by denying that he had her authority. The Queen, as usual, would not commit herself. Though

S. P. Spain, xxvi. The confusion of erasures, corrections, and interlineations, where Drake's case is dealt with, would afford fairly clear evidence of the complicity of the Government, were no other to be had.

she smiled and held the rod behind her back, it did not suit her diplomacy just then to disown the blow with which Philip's extremities were tingling. It was just what she wanted, to whip him into the ratification of Alva's convention. So it was only a question of time for Drake to be able to reappear out of hiding, and hold up his head without fear of the Lord Admiral, and in the proud consciousness of an exploit that was dear in the eyes of his sovereign.

Nor did he doubt it was as dear in the sight of Heaven. As Elizabeth consulted Jewell, so Drake had consulted his spiritual adviser, and the parson had told the seaman, as the bishop told the Queen, that to prey on the idolaters was doing God a service. We may laugh in cynical distrust of such convenient doctrine, and doubt the tenderness of consciences so simply satisfied; but it was all real and sincere enough then. The Elizabethan Protestants went to the Bible for their political morality as a man goes now to his newspaper; and in the pages of Scripture they found writ large and clear a command for unceasing war on idolatry. Nothing was plainer to men like Drake than that the Mass was an idolatrous rite. He had seen the Spaniard abasing himself beside the passing Host: he had seen the African negro grovelling before his Mumbo Jumbo; and too simple-minded to grasp the higher mysticism of the Catholic creed, he could find no difference between the two states of mind. So with all the devotion of Gideon he warred upon the idolaters, and revelled like a Hebrew captain in the spoil of the heathen. It was to him a crusade; and like a crusader he made war. Never once was his creed made to serve as a cloak for piratical excesses. For him his religion was as much a bridle as a spur. Implacable as was his animosity against Spain, Spaniards were universally won by the courtesy and even the generosity with which he treated them. He never killed a prisoner, no matter what he risked by sparing his life. He never destroyed a ship unless it were taken in act of war against him. His hate was heroic, and he fought his enemies as though he loved them. For a church, or a woman, or an unarmed man, he had a noble forbearance that puts the brightest chivalry of his time to blush, and it was the grateful eulogies of his prisoners of war that crowned his reputation.

Still, in spite of all his chivalry, politicians who believed his exploits were dangerous would not be persuaded. These men regarded rivalry with Spain as madness. They were content if England maintained her position as a second-rate power, and picked up a subsistence in such corners of the earth as she could find unoccupied by Spain. Thus, though the law did not lay hands on him, he was not permitted to put in practice the scheme with which his heart was aching, and with admirable patience he submitted to the restraint. His passion never ceased to consume him. Yet month after month, and year after year, he waited for the Queen to sanction his quest. His only solace was to send his brothers and companions to continue his work, and to watch cruiser after cruiser leaving Plymouth for the West. Fired by his success, Hawkins, Wynter, and half the Devon gentry were fitting out vessels to follow where he had led, but Drake remained at home.[1] Poor

[1] *Memorias de los Corsarios*, etc., *ubi supra.*

John Oxenham, who had vowed to be his fellow in the great enterprise, was not so patient. After waiting three years, he stole away to the South Sea overland, and being taken by Spaniards, with no commission to show, was hanged for a pirate on Lima gallows, as every one knows.

More wise and loyal than his lieutenant, Drake sought relief in the public service. In August, 1573, a few days after his return, the Earl of Essex had gone to bury his reputation in Ulster. In view of the coming struggle with Spain, Ireland was as great an anxiety for Elizabeth as the Low Countries were for Philip; and the chivalrous Earl had craved permission to undertake the quest, and to reduce the island to his mistress's obedience with his own Quixotic lance. Drake was there too, if legend says true, hiding in Queenstown harbour, where in the creek that still bears his name he was lost to his persecutors and Philip's cruisers.[1] There he lay till his pardon was sure, but with his danger faded all hope of a venture to the Pacific. As the year 1574 grew old the Government drew closer and closer to Spain. Walsingham and Leicester were still for defying Philip, and openly assisting the Prince of Orange; but Burleigh and Bacon had gone over to the party of the Spanish alliance. Alva was now at Philip's ear, and Spain was as effusive as England. The English refugee traitors were banished from Philip's dominions; and Sir Henry Cobham went over to Madrid to negotiate a commercial treaty. Thus, when at last Drake was able to emerge from his lair, he found the air filled with such a peal of harmony

[1] The tradition of his hiding in Drake's Pool can only refer to this period.

HOFSTRA UNIVERSITY
LIBRARY

that it was useless to expect his sighs to be heard. In despair he resolved to seek service in Ireland, and in the spring of 1575, armed with a letter of introduction from Hawkins, he joined the Earl with his frigate the *Falcon.*[1]

Around Essex was gathered the flower of English knight-errantry, and the adventurous seaman was received with open arms into their company. Here were Black John Norreys and his brother William, and others of their stamp fresh from the Low Country wars. They were the last of the old race of medieval soldiers whom the Prince of Orange's mathematics were soon to improve away—"breathing valiant" men who, hot of head and stout of heart, gave to Shakespeare his stormy captains, and like them bigoted, quarrelsome, and loyal, loving, hating, and fighting, raged through their lives at Homeric pitch. These, at an age when a man's nature receives its last impressions, were Drake's constant comrades; with these men he now for months shared danger and privation; what wonder if the strong fellowship that such an atmosphere alone engenders left its mark for ever on the adventurous sailor?

Nor was this the sum of his changed surroundings.

[1] *Harl. MSS.* 540, fol. 103, *b.* *S. P. Ireland,* liii. 49. *S. P. Ireland* (Folios) vol. viii. Stowe says, "he furnished at his own proper expense *three* frigates, with men and munitions, and served voluntary." Essex speaks of having three frigates in his service, but calls them "your Majesty's frigates" (Essex to the Queen, Devereux, i. 113-117), and the accounts show Drake receiving pay as captain of the *Falcon,* and not otherwise. This, however, appears to be the first introduction of this class of vessel into the English service, and Drake may have presented them to her—to shut the Lord Admiral's mouth on the subject of piracy.

If the age still bred its Hotspurs, it was begetting Iagos too, and such a one was Mr. Thomas Doughty. Of all Drake's comrades, this accomplished gentleman won the largest share of his affection. He was a man whose nature, once perhaps admirable, had been poisoned by the atmosphere of intrigue in which he had lived. It was a type which more justly reflects the age of Elizabeth than any one of those brilliant figures who by the very fact of rising above the ordinary level most attract the attention, and it was a type most nicely calculated to win the rougher nature of Drake. He was a scholar of no mean pretensions, and could display both Greek and Hebrew; he had served a campaign or so in the Low Countries, and gracefully supported the reputation of a soldier; he had studied law at the Temple, and could discourse in honeyed phrases the fashionable philosophising of the hour. Drake always loved a scholar as he loved a soldier, especially if he were a Cambridge man, as Doughty seems to have been. Even Essex had been won by his parts. He had been in a confidential capacity in the Earl's household, and when Essex found his work was being thwarted, Doughty had been sent over to try and remove the obstruction at Court. But he was now in disgrace, for he had brought back a lying report that the difficulties were all due to Leicester's slanders. Essex had written a furious letter of complaint. Leicester had explained, and Essex in a dignified apology declared that he should withdraw his confidence from Doughty. The discredited servant continued, however, to serve as a soldier. Drake and he became inseparable, and so brotherly in their affection that the seaman even imparted to his friend the secret on which his whole heart was bent, and they vowed to unite their efforts in bringing the great adventure into being.

It is easy to understand the delight which Drake's humanity found in the polished society of such a man. For the warfare in which he was engaged was of a fierce brutality beyond anything he had seen. The Irish seas were swarming with pirates, and in burning their galleys and supporting Essex in his wretched man-hunts, the frigates were chiefly engaged. All was murder, rapine, and fire, and the piteous spectacle culminated in the last act of Essex's proconsulship. To the isle of Rathlin the chiefs of the rebels and the invading Scots had sent their women and children for refuge, and so heartless was war in those days that even this pattern of Elizabethan chivalry conceived the idea of destroying them all. As the Earl retreated to the Pale, John Norreys with his company was left behind at Carrickfergus under orders to concert with the sea-captains the surprise of the sanctuary. One day in July, a flotilla escorted by the three frigates suddenly left the harbour. Two days later, in spite of every difficulty, a landing was effected. The first assault on Bruce's Castle was repulsed, but on Drake and his fellow-captains getting two heavy guns ashore, the Scots leader speedily capitulated. The garrison was given over by Norreys to the vengeance of the soldiers; two hundred souls were massacred as they left the castle, and then day after day a cruel hunt went on till every cave and hollow of those storm-beaten cliffs had echoed with the victims' shrieks, and not a soul—man, woman, or child—could be found alive in St. Columba's Isle. So miserably did those two famous captains first join their hands in war. We can but turn from the spectacle in disgust, and try to think of it as the parent of that most chivalrous venture when Drake and Norreys went out alone to fight the nation's battle, and set an exiled monarch on his throne. Drake himself, while the massacre went on, was busy with the frigates, burning eleven Scottish galleys; and had it not been so, still we could hardly blame him for sharing an exploit which the Faery Queen's own knight could describe to her in an exultant despatch.

With the close of Essex's mission Drake's services were no longer required in Ireland, and he came home with a glowing letter of recommendation from the Earl to Walsingham. Essex had been thoroughly impressed with the capacity of his new officer, and now that he could no longer find him work he sent him to the spirited Secretary of State as a man fit to serve against the Spaniards. The expression need not surprise us. Through good report and evil report Walsingham had been Essex's staunchest friend. Both were consistent supporters of the war policy, and Walsingham may even have asked for a likely man; for once more the wind had gone round and was blowing gustily from the stormy quarter. Cobham's mission had proved a failure, and he had returned without having removed the Grand Inquisitor's opposition to the proposed exterritoriality of heretic merchantmen. In Don John of Austria a new champion for Mary Stuart was sounding his challenge in the lists, and under his countenance the English Catholic refugees had fitted out a pirate fleet. Daily the Council were debating a war with Spain, and seriously considering the proffered allegiance of Philip's Dutch rebels. Walsingham was now supported by all the Council except the old peers and Philip's pensioner. Even the cautious Burleigh was catching the war-fever. The city merchants, trembling for their trade, were still hesitating, when, in January, 1576, came the news that a ship of Sir Thomas Osborne, one of the greatest of the merchant princes, had been seized in a Spanish port, and its crew flung into the dungeons of the Inquisition. It was the last blow to the time-honoured alliance, as the affair of Vera Cruz had been the first. The country was thoroughly roused. Elizabeth was furious. Parliament was summoned to provide the sinews of war, and Cobham went over to Brussels to threaten that if Alva's convention were not at once ratified, the Queen would put into execution "some remedy for her relief that she would not willingly yield unto."

It was no idle threat. Soon after the delivery of Essex's letter, Drake had been surprised by the sight of Walsingham's grim face in his lodgings. As soon as they were alone the Secretary had unfolded a map, and informing Drake that the Queen had received injuries of the King of Spain, for which she desired revenge, asked him to mark upon it the points at which Philip might be most annoyed. Thinking his hour had indeed come, yet cautious still in the midst of his excitement, Drake began vaguely to hint at his mighty project. Walsingham at once asked him to reduce it to writing and sign it. But the wary seaman was not to be caught. No doubt he preferred to keep his own secret, for the King of Spain's eyes might be anywhere; besides, he was too good a Protestant not to be overflowing with the idea of which the assembling Parliament was full, and he refused to put pen to paper. "Her Majesty is mortal," he told the Secretary, "and if it should please God to take Her Majesty away, it might be that some Prince might reign that might be in league with the King of Spain, and then will mine own hand be a witness against myself." It was all Walsingham could get out of him then. But a few nights later he received a summons from the palace, and was introduced by the Secretary into the presence of the great Queen herself. With all the witchery she so strangely exercised over the sturdy chivalry of her time, she appealed to the young sea-captain as some distressed princess to an errant knight. Her caitiff kinsman had foully wronged her, she was pining for revenge, and he alone was worthy of the quest. So with a woman's art she claimed his service and drew from the eager seaman the outlines of his immortal project for a raid into the Golden Sea.

Full of hopeful resolve he left her, but it was only for new disappointment. By the end of March the wind had changed again. The wayward Queen had quarrelled with her Parliament for being as Protestant as Drake, and had rudely dismissed the Dutch envoys. The breach between the two natural allies widened as the summer went on, till in the autumn Elizabeth was actually contemplating an active policy against the Prince of Orange. Walsingham was in despair. He saw his efforts to open the perverse Princess's eyes were useless; he saw she had obstinately made up her mind that Philip meant peace, and determined to save her from herself, his unrivalled genius for management seems to have shown him in Drake an instrument to force her into war with Spain.

How the affair was managed it is still impossible to

say; but there can be no doubt that the Queen was approached through Hatton, and no one at that moment was better calculated to lead her into mischief than her handsome captain of the guard. Fortunately Thomas Doughty's talents had again served him in good stead, and he was now Hatton's private secretary. Doughty's eloquent tongue turned the favourite's head with the dazzling prospect which the venture promised, and in due course the Queen told Drake she was ready to subscribe a thousand crowns to the expense of an expedition to penetrate the South Sea; but it was on the express condition that the whole affair was to be kept a profound secret, and she gave stringent orders that above all not a word of it should be breathed to Lord Burleigh.

So like a naughty child did this strange woman try to cheat her most trusted counsellor. But the Lord Treasurer was not so easily outwitted. He knew everything, though he held his peace, and quietly went to work to prevent the prank on which his wayward mistress was bent. Sad to relate, a tool was ready to his hand. No one knew better than he an Iago when he saw him, and with such a man he could play as he would. Doughty, ignorant perhaps of Drake's relations with Walsingham and the Queen, felt that his own Court influence had done everything, and jealous of the prominent position which Drake assumed, revealed the whole project to Burleigh. It takes away the breath to think that the great Minister with whom all the glories of Elizabeth are associated should have deliberately set to work to wreck the most brilliant and typical achievement of her reign; yet what the instructions were which Doughty received from the astute old schemer became only too plain in the sequel.

Meanwhile Drake, without a suspicion of his friend's treachery, threw himself into the work of floating his company and organising the expedition. It was in a happy hour, for just now the town was run mad on exploration. In the autumn of 1576 Frobisher returned from his attempt to discover a north-west passage into the South Sea, and had turned every one's head with some ore which he had brought from Labrador. Court and city vied with one another in supporting his venture for 1577, and Drake caught the tide of speculation at the flood. Leicester, Hatton, Walsingham, and other courtiers took shares freely. Sir William Wynter, Elizabeth's Admiral-at-Sea, so warmly supported the scheme that the younger Wynter was appointed vice-admiral. It also seems to have received the sanction of the Hawkinses, for young William Hawkins, the son of the Governor of Plymouth, sailed with it. Cadets of the best Devonshire families freely volunteered, and when the little sea-captain with his fine clothes and his page, and Diego, his negro, strutting at his heels, swaggered into the Temple to see Tom Doughty, there were found plenty of spirited young barristers weary of their briefless existence and eager to embark their younger sons' portion in a romantic adventure. Drake's frank blue eyes and hearty self-reliance spread confidence around him, and he was soon busy equipping his expedition regardless of cost.

In Plymouth harbour three ships were brought forward for a voyage to Egypt. There was the *Pelican*, of one hundred tons and eighteen guns, which was to serve him for his "admiral" or flagship; the *Marygold*, a barque of thirty tons and sixteen guns; and a provision-ship of fifty tons, which he called the *Swan*, after the little vessel in which he had founded his fortunes. In September John Wynter came round from the Thames with a fine new ship of eighty tons and sixteen guns, named the *Elizabeth*, and a pinnace of fifteen tons, which Drake re-christened the *Christopher*, in honour of Hatton. The *Marygold* and the *Swan* were commanded by two of the gentlemen volunteers, Mr. John Thomas and Mr. John Chester; while the *Christopher* was committed to Tom Moone, the trusty carpenter who had scuttled the old *Swan* off Cartagena five years ago. Doughty seems to have considered himself as Captain of the Land-soldiers, but otherwise had no command. The crews, all told, numbered one hundred and fifty men and fourteen boys, and included men of science as well as combatants and mariners. Amongst these Doughty was careful to enrol an ally. For some months past his brother John had been confined by order of the Council without a trial, on a charge, there is reason to believe, not unconnected with the scandal about Leicester having poisoned Essex. Hoping probably never to hear of him again, Leicester had been recently induced to consent to his release, and for this ominous recruit Doughty obtained a place as gentleman volunteer.

From a natural lavishness, no less than a sagacious belief in the efficacy of display, no expense was spared to make the squadron as splendid as possible. Skilful musicians were engaged, and arms and munitions of the best patterns provided, while the flagship was fitted with all the splendour proper to the dignity of its commander's commission as Captain-General. His table furniture was of silver, richly gilt and engraved with the family arms; his cabin was redolent with perfumes, which the Queen herself had given him; and in every way he was surrounded with all the state and luxury of one of her Majesty's Admirals-at-the-Sea.

To the end Drake suspected nothing. True, Doughty had been overheard in Drake's garden at Plymouth making use of expressions which seemed to hint that the Queen and the Council were to be bribed into winking at piracy; but Drake would hear nothing against his friend. Indeed there is no reason to believe that Doughty revealed the destination of the squadron to the Spanish agent, though the Spaniard must have guessed that the thin disguise of a venture to Alexandria covered another of those sharp surprises with which Elizabeth loved to season her diplomacy. Don John had assumed the government of the Spanish Netherlands, and the exiled English rebels were gathering round him in spite of Alva's convention. In the summer letters were intercepted revealing his romantic dream of conquering England for the love of Mary Stuart; and the Spanish agent, caught in correspondence with the Scots Queen, was sent to the Tower. At last Walsingham and Leicester persuaded Elizabeth she was being betrayed, and war was on every one's lips. Arms and troops and projects to annoy the King of Spain were the only talk in the Council; and it is impossible to believe that secret orders had not enlarged Drake's innocent commission for trade and exploration, as on November 15th, 1577, at the height of the war-fever, he ran proudly out of Plymouth harbour.

CHAPTER V

AN OCEAN TRAGEDY

In days when the earth seems hardly to have a secret still unprofaned, it is difficult to grasp through what a world of shadowy terrors Drake had resolved to push his way. Yet we may conceive the strange fancies that mixed with the music of wind and wave, of trumpets and violins, as he paced the poop of his little flagship and watched her four frail consorts rising and falling to the mysterious swell of the Atlantic. Since the discovery of the New World no event had so profoundly moved men's imagination as Magellan's great achievement sixty years ago. Till then it was believed that America was part of one vast continent that covered all the South Pole, and was one with New Guinea; and even now, geographers taught that there was no southward passage from the Atlantic into the South Sea except by the narrow strait through which the great Portuguese discoverer had passed to find the Old World beyond the New. Time after time the most renowned officers in the Spanish service had attempted to follow in Magellan's track, but even those who succeeded in passing the strait had demonstrated with disaster the impracticability of the route. For a generation it had been abandoned: the riches of the South Sea continued to pass overland by Nombre de Dios; and primeval silence had once more settled upon the desolate shores of Patagonia.

To the real and known difficulties of the navigation, the science of the day added all the terrors of its naïve deductions. The *primum mobile* was supposed to set up so violent a westerly current that even if a man passed in he could never return. So great was the dread the idea inspired that Magellan had been compelled to hang two of his chief officers before he could proceed, and the last attempt to penetrate the passage had been frustrated by open mutiny. For those who had no science, myth and legend provided horrors in plenty. The seafaring folk of Europe generally believed the fires of Heaven would consume all who attempted to pass the torrid zone; and those whose comrades had passed and lived, still shuddered at a void beyond where, engendered of incessant storm and darkness, the great Atlantic swell was born. Yet it was through this region of prodigies, chartless and unexplored, that Drake with his imperfect instruments was going to find his way; and these were the men whom, with no help of modern discipline, he had to persuade to the work. They had only agreed for a voyage to Alexandria, nor was it till the little squadron rendezvoused at Mogadore, on the west coast of Morocco, that they knew anything more adventurous was in the wind. At any time they might refuse to venture farther into the burning void. With all the support of his officers Drake could barely hope to inspire them with his own daring; and at his side was his dearest friend sworn to overthrow the voyage.

It was not long before Lord Burleigh's tool was at work. Running down the African coast as far as Cape

Blanc, Drake again put in with half-a-dozen prizes he had picked up on the way. Here he proceeded to clean his ships, and while the rummaging was going on Doughty got the soldiers ashore to exercise them at their weapons. There was a prospect of sharp work ahead, for Drake meant to water and victual at the Cape Verde Islands, and force might be needed. Mayo, which was then a notable haunt of pirates, was the first point attempted, and here Doughty, with Captain Wynter in his company and some seventy musketeers, was landed to scour the island for provisions. Once out of sight, he began secretly tampering with the men, and after a half-hearted attempt to trade with the inhabitants returned to the ships with a report that nothing was to be had. It is possible that this unsatisfactory performance already aroused Drake's suspicions, but the constancy of his friendship suffered him to give no sign; and when next day off St. Iago a rich Portuguese prize bound for the Brazils was captured, Doughty was placed in command. Besides silks and other valuable goods, it contained all the stores which the fleet needed, with the exception of such things as were to be had in abundance at the island of Brava. Thither, therefore, Drake at once proceeded, taking no further precautions with the prize than to send his brother aboard to represent his interests. Thomas Drake seems to have been the last survivor of the eleven sturdy Protestants who had been reared on the Medway hulk; and, with the exception of Tom Moone, he was perhaps the only man in the expedition on whom the Admiral could implicitly rely. Arrived at Brava, Drake went aboard the prize to arrange for the discharge of the prisoners, and found himself at once face to face with his new difficulty. No sooner did he step on deck than Doughty came up and accused Thomas Drake of having pilfered the cargo. On inquiry, one of Hatton's men, who was now Drake's trumpeter, made a similar charge against Doughty, and not only was his accusation supported by others, but some trifling articles belonging to the prisoners were found in Doughty's possession. In vain he protested they were presents. Where prize was concerned such a defence was inadmissible, and Drake flew into a passion. He told his friend plainly that it was not Thomas Drake, but Francis, that he meant to disparage; he accused him of seeking to sap his credit with the fleet, and swore by God's life he would not have it. For, unlike the later Puritans, Drake was no precisian. He could swear like a gentleman, and on his occasions used his fashionable accomplishment with some freedom. Still it was not his way to bark without biting, and he ended his outburst by ordering Doughty aboard the flagship, and giving the command of the prize to his brother.

With the exception of one smack which had been exchanged for the *Christopher* pinnace, the whole of the other captures had been already dismissed; but the Brazil ship so nicely suited Drake's needs that he had resolved to attach it to the squadron as an extra victualler, and before leaving the confines of the Old World, where now they were, he set at liberty all his prisoners. Drake's gentlemen were not a little astonished at his clemency; it was not at all their idea of honourable piracy. Some of the Portuguese passengers were men of wealth and position; yet not only were they discharged without ransom, but a newly-set-up pinnace was given them that they might get back comfortably to St. Iago. One, however, remained behind, and the exception is important. He was a pilot for the Brazils from Genoa, with a Genoese love of adventure; and so soon as he heard Drake's intention of penetrating into the Pacific by Magellan's abandoned route, he volunteered for the voyage. His services were accepted gladly enough, and with this valuable addition to the staff, the course was laid for the River Plate.

For the present no further notice was taken of Doughty's misconduct. Mr. Vicary, one of the barristers who had volunteered from the Temple, interceded in his behalf so successfully that Drake left him in peace on board the *Pelican*, while he himself sailed in the prize. But his forbearance was of little avail. As they crept across the Doldrums, constant complaints came from the *Pelican* that Doughty was trying to take upon himself the command of the ship. The gravity of the situation lay in the fact that the position of soldiers and gentlemen volunteers on board ship had not yet been determined, and the consequent jealousy and insubordination were exactly the instruments best suited to Doughty's hand. It was a question of extreme delicacy, and on its right adjustment the efficiency of the navy hung. At Brava Doughty had begun by assembling the *Pelican's* crew, and while charging them to look to the master in all matters of seamanship and navigation, he had given them to understand that he was there as Drake's deputy to exercise the powers of life and death contained in the Queen's commission.[1] In the absence of the captain,

[1] See "The sum of Thos. Doughtie's oration upon the *Pelican*, etc.," *Harl. MSS.* 6221, *fol.* 7, omitted in Vaux's collection, *Hak. Soc.* 1854.

Doughty may really have thought that he was entitled to command. Lieutenants had not then been introduced into the sea-service. The second officer of the ship was the master, invariably a practical seaman; but in the Spanish service seamen had always to give place to soldiers, and as Spain then set the fashion in all things military, Doughty probably thought it a duty he owed to himself and the other gentlemen to teach the master his place. But Drake knew better, and face to face with the question for the first time he grappled it with all his masterful directness. No greater debt is due to him than that he successfully resisted the ideas which paralysed the Spanish navy; and it is in that very arrogance which disgusted so many of his contemporaries, that much of his greatness must be sought. For from it was bred that blustering pride in his profession which for the first time taught soldiers to respect their brethren of the sea; it was his spirit that inspired Shakespeare's Boatswain; it was he who made "Out of the way, I say!" the standing order for soldiers aboard English ships; and it is to the high credit of both teacher and taught that no admiral was ever more popular with the military than Francis Drake.

As Doughty had been sent aboard the *Pelican* under reprimand, his case was one of peculiar aggravation, and things soon came to a crisis. The offender never ceased his pretensions, and one day so far forgot himself as to take advantage of some rough practical joke that the seamen were playing on the Admiral's trumpeter, to be revenged on his accuser. The man had apparently gone on board the flagship in the course of his duty, and

Drake seems to have regarded the insult as a piece of deliberate insubordination. On the trumpeter's complaint, he sent a boat for Doughty, and so soon as he came alongside, without permitting him to set foot on the ship or to say a word in his defence, he peremptorily ordered him on board the victual-ship in utter disgrace.

On April 5th the coast of Brazil was made about Rio Grande, and here they lost touch of the fair weather that had attended them. Sudden fogs, accompanied with heavy weather, scattered the ships and drove them one after another deep into the mouth of the La Plata in search of water and shelter. It was not till the end of the month that they all got together again, and Drake, resuming his command of the *Pelican*, ventured to continue his course to the south with his reassembled fleet; but no sooner had he doubled Point Piedras than another storm struck them, and the victualler parted company. Doughty was still on board of her, and Drake, who like other sailors had his superstitions, began to think his Jonahs were brewing the bad weather. They had never ceased to tamper with the men, and in his anxiety to discredit the Admiral and advance his brother's party, John Doughty had used his Hebrew and Greek to claim acquaintance with the black art. The incessant gales which Drake encountered as he struggled southward in search of a port to reorganise the squadron did little to remove his suspicions. The ships were continually losing touch, and so sure as they attempted to ride, a gale would tear them from their anchors. Did Drake venture inshore with his pinnace to explore the coast, a squall would strike him so suddenly that only by the daring seamanship of his officers could he be rescued from destruction. His brother in the prize was missing altogether; and the victualler, with its uncanny passenger, had not been seen since it first parted company at Point Piedras.

For six weary weeks the struggle southward continued till five degrees north of Magellan's passage Port Desire was discovered. Here the storm-beaten fleet found rest; and now at the moment when all danger seemed over, the victualler mysteriously reappeared to enjoy the security. But this was not all. Her master had to report that not only had Doughty never ceased to disparage the Admiral and make himself appear as the real commander of the expedition, but that he had done all in his power to get Mr. Chester, the gentleman captain of the victualler, to quarrel with the master and defy his authority. So serious had the situation on board become that some one had even gone so far as to remind the offender of the fate of Magellan's mutinous vice-admiral, but Doughty had only laughed and said Drake had no authority of life and death, and gallows were for dogs, not gentlemen. Even if Drake's power of life and death was ever intended to apply to the officers and gentlemen of the fleet, it is highly improbable that he at all contemplated such an extremity. But it is certain that he was bent on upholding the authority of the sea-officers, and that Doughty, with Burleigh's instructions and his own end in view, was deliberately fomenting the jealousy between them and the gentlemen. Still Drake seems not to have despaired of bringing his friend to reason. Being in want of firewood, and desirous of making his squadron more compact, he resolved to break up the victualler, of which there was no longer need; and while the work was going on he once more took Doughty on his own ship. But so persistently did he continue his efforts to paralyse the expedition, that one day in a fit of exasperation Drake ordered him to be bound to the mast. It was an ignominious sea-punishment, well designed to teach the gentlemen their place. But the matter did not end there. The continued friction was fast chafing Drake's masterful spirit to a dangerous heat, and as soon as the offender was released both he and his brother were ordered on board the smack *Christopher*. At last the conspirators began to be alarmed. It was Tom Moone's ship, and the trusty old carpenter was the very pattern of a pirate's lieutenant. Truculent, fearless, and devoted, he was Drake's chosen instrument for deeds he dare not own; and reading murder in the grim seaman's eyes, the brothers refused to obey Drake's order. His only reply was to direct tackle to be rigged to sling them on board.

On June 3rd the four vessels that were still together again stood southward, hoping to find that Thomas Drake's lost vessel had preceded them. Again they encountered adverse gales, again the vessel which carried the Doughties parted company, and again, after Drake had desisted from the hopeless struggle southward and was running back up the coast, she reappeared. He was now convinced that the weather was due to sorcery, and, determined to make his squadron still more compact, he resolved to abandon the *Christopher*. Tom Moone was taken on board the *Pelican*, and the Doughties were handed over to Captain Wynter on the *Elizabeth*, with strict orders that no one should speak to them, and that neither of them should be allowed to read or write anything but what a man could see and understand. It is easy to smile at such credulity; but rather should we bow before the undaunted spirit which, oppressed as it was with imagined terrors, could yet bravely lift the load of opposition which each day grew more real. For now the crisis of the voyage was at hand. The harassed Admiral had given the order to stand onward as far south as the latitude of Magellan's Straits; every man in the fleet knew at last for what he had been brought so far, and at any time Drake might find mutiny staring him in the face. The least credulous might well believe that they were already in the confines of that fabulous storm-land, in which Thomas Drake and their comrades in the prize seemed already engulfed. As they knew the world, it was summer-time; and yet, as they painfully beat southward, at every league the skies grew more wintry and the sea more tempestuous, till with infinite toil having reached the required latitude, for the third time they were hurled back.

It was an ill wind, but as though Drake's precautions had paralysed the Doughties' magic, it blew him to his brother. So severely, however, was the prize found to have suffered in the storms that Drake resolved to put in at Port St. Julian to finally refit for the desperate attempt. It was the natural harbour where sixty years ago Magellan had made his last preparations; and well-nigh overwhelmed with fatigue and anxiety, Drake entered it in safety on June 20th. It was like the end of the earth. For six months he had been sailing out into a world on which God's back seemed turned. Yet there it was, upon the shore of that forsaken wilderness, that the first sign of Christian men fell upon his eyes; and that sign was the stump of Magellan's gallows.

What wonder if, as Drake with troubled brow gazed upon that jagged fir-post, the ghost of the old admiral's resolution whispered in his ear, and he saw amidst the desolation a sign from Heaven? Buried at its foot they found the skeletons of the two mutinous officers, while on board the *Elizabeth* the presence of the prisoners was fast demoralising Wynter's ship's company. It was clear the situation could not continue, though Doughty was still confident that the Admiral dare not exercise his powers upon a gentleman. But he knew not his friend, nor could he measure the spirit to which the formulas that bind the world in chains are but as threads. Drake was no man to suffer a great purpose to be strangled with the phrases of a parchment. By the old law of England an offender could be always condemned by the judgment of his peers, and so by first principles he cut the knot. On the last day of June the crews were summoned ashore, and there over against the gallows of Magellan Drake sat in judgment upon his dearest friend. The time-hallowed forms of the English law were reverently preserved; a jury was empanelled with Wynter at its head, and solemn articles were read which charged the prisoner where he stood with mutiny and treason. Then with bitter taunts and acrimonious evidence the wrangling trial went on. Hearsay, prejudice, and abuse were heaped on the wretched prisoner's head in a way that shocks us now to read. But such was then the everyday scene in the old courts of England where our liberties were shaped; and few prisoners fared so well at Westminster for another century as did Thomas Doughty at that first Lynch-court amidst the desolation of Patagonia. Though every one believed him guilty of treason, he was acquitted because the evidence was insufficient—an unusual piece of clemency in days when juries were expected to convict on their general impressions. It was not till they had found him guilty of mutiny that Drake produced any evidence himself. At the beginning of the trial he had protested that it was no matter of life and death; but in the midst of one of the sorry wranglings Doughty boasted of having betrayed the Queen's secret to Lord Burleigh. Then at last the whole truth burst upon Drake, and he knew the case was even more desperate than he thought. Leaving the traitor standing alone with his brother, he called the companies down to the shore, and laid his heart bare to them. He told them the whole story of the expedition from first to last, he told them what it meant, and asked them what a man deserved who had conspired to overthrow so great an undertaking. "They that think this man worthy of death," he cried out at the last, "let them with me hold up their hands." And as the words left his lips a throng of brown hands surrounded him.

On the second day from the trial the tragedy was played out. Wynter, it is said, made an effort to save the culprit, but Drake's hand was set firm upon the plough. On an island over against the relic of Magellan the block was placed, and beside it an altar where side by side the two friends knelt to receive together the Sacrament in token of forgiveness. Hard by, tables were spread with the best the stores provided, and there they all caroused together in a farewell banquet to their comrade. When the feast was ended, with courtly jests Doughty drew near the block and the boon companions gathered round. At the last, as one who had lost in a game of hazard, he embraced the friend who had won, and Drake took payment without a flinch. He showed no animus, nor did sentiment sap his purpose one jot. Like everything else, his affection had to be sacrificed to the mission. Doughty had stood in the way of the great lesson he meant to teach his country, and he had been honourably removed. That was all. So the sword fell, and when the provost-marshal held up the dripping head, Drake cried out, unmoved, "Lo! this is the end of traitors."

What wonder if his heart was hardened? What wonder if it was said of him afterwards that he was a man hard to reconcile? Such a tragedy might well have poisoned altogether a nature less magnanimous. He would always speak of his friend with love and admiration; but the memory of his treason never failed to rouse in Drake something that made him terrible to his officers.

John Doughty he spared with a wise clemency; but the rest had still to be taught the lesson which the dead man had so hardly learnt. As the work of cleaning and refitting went on, the tension between the gentlemen and the sailors continued, till Drake could endure it no longer. He was overwrought and desperate with the troubles that beset him. A month went by, and then once more he called the companies ashore. The chaplain thought it was for a sermon, but Drake said he would preach himself that day, and a wholesome sermon it was. He told them the mutinous discords must cease or the voyage would be overthrown. "I must have the gentleman to haul and draw with the mariner," he cried, "and the mariner with the gentleman. I would know him that would refuse to set his hand to a rope." He offered the *Marygold* to any who wished to go back. "But let them take heed," he said, "that they go homeward. For if I find them in my way I will surely sink them." With one accord they all consented to go on and leave the wages to him. Then turning to Captain Wynter, who stood at his side, he dismissed him his ship, and cashiered every officer in the squadron. They asked him why. "Is there any reason why I should not?" he retorted. As he grew more excited he rounded upon the traitors that he knew, and in terror they humbled themselves at his feet. So in triumph he told them once more how it was the Queen who had sent him out, and how they had come to set the kings of the earth by the ears, and warned them of the fate that awaited them if the voyage turned out a failure. With that, as suddenly as before, he restored every one to the rank of which he had just been deprived, and with cheery words of hope and kindliness he dismissed them to their duty.

From that moment his reputation as a disciplinarian was unrivalled. The state of his ships was a wonder to all who saw them, and Spaniards themselves considered his men as comparable only to their own Italian legions. No more was heard of the quarrels and jealousy. The work went rapidly on: the prize had been broken up to supply the other three ships with firewood; and on August 20th, 1578, the three ships that remained hove-to before the Straits of Magellan.

CHAPTER VI

WAKING THE SOUTH SEA

Lord Burleigh's scheme had failed, and Drake was knocking at the golden gates. In the teeth of the astutest Ministers of the time, he was about to blow the blast before which the giant's doors would fly open, and deliberately to goad the giant into open fight. Full of the momentous meaning of his resolve, he paused upon the threshold to do honour to the mistress whose favour he wore. Before the frowning entry he caused his fleet, in homage of their sovereign lady, to strike their top-sails upon the bunt as a token of his willing and glad mind, and to show his dutiful obedience to Her Highness. It was a piece of true Elizabethan chivalry, and like a true Elizabethan knight he accompanied it with a shrewd stroke of policy. Sir Christopher Hatton had now no visible connection with the venture. The vessel named after him had been broken up, and his representative had been beheaded. Drake knew well how flat fell prowess at the Faery Queen's Court if a man had not a friend at her ear. He knew, too, that no reputation was so fashionable just then as that of a patron of discoveries, nor could he be ignorant that all the new favourite's goodwill would be required to save him from Burleigh's power. So on the poop of the little flagship was placed the crest of the Captain of the Guard, and in his honour the *Pelican* became the *Golden Hind.*

So protected, Drake boldly entered the Straits. Then from the towering snow-cones and threatening glaciers that guarded the entry the tempests swept down upon the daring intruders. Out of the tortuous gulfs that through the bowels of the fabulous Austral continent seemed to lead beyond the confines of the world, rude squalls buffeted them this way and that, and currents, the like of which no man had seen, made as though they would dash them to pieces in the fathomless depths where no cable would reach. Fires lit by natives on the desolate shores as the strangers struggled by, added the terrors of unknown magic. But Drake's fortitude and consummate seamanship triumphed over all, and in a fortnight he brought his ill-sailing ships in triumph out upon the Pacific. Then, as though maddened to see how the adventurers had braved every effort to destroy them, the whole fury of the fiends that guarded the South Sea's slumber rushed howling upon them. Hardly had the squadron turned northward than a terrific gale struck it and hurled it back. The sky was darkened, and the bowels of the earth seemed to have burst, and for nearly two months they were driven under bare poles to and fro without rest in latitudes where no ship had ever sailed. On the maps the great Austral continent was marked, but they found in its place an enchanted void, where wind and water, and ice and darkness, seemed to make incessant war. After three weeks' strife, the *Marygold* went down with all hands; and in another week Wynter lost heart, and finding himself at the mouth of the Straits, went home in despair; while the *Golden Hind*, ignorant of the desertion, was swept once more to the south of Cape Horn. Here, on the fifty-third day of its fury, the storm ceased exhausted, and Drake found himself alone. But it was no moment to repine, for he knew he had made a discovery so brilliant as to deprive even Magellan's of its radiance. He was anchored among islands southward of anything known to geographers, and before him the Atlantic and Pacific rolled together in one great flood. In his exultation he landed on the farthest island, and walking alone with his instruments to its end, he laid himself down, and with his arms embraced the southernmost point of the known world.[1]

The spell seemed broken. As when the hero blows the magic trumpet, the dreaded bulwarks had fallen, the brazen gates of the castle had disappeared, and in their place stretched a broad and smiling way into the golden garden. The very forces of Nature that had seemed a part of the enchantment ceased their struggle to stop the man that made such light of their terrors, and let him go in triumph on his path of plunder up the slumbering coast of Chili.

About a month later, little dreaming what had taken place, the crew of the *Grand Captain of the South* were lazily waiting in Valparaiso harbour for a wind to carry them to Panama with their cargo of gold and Chili wine.

[1] *Observations of Sir Richard Hawkins*, p. 224, Hak. Soc. 1878. Drake's discovery first appears on a Dutch silver plaque executed in his honour, apparently, during his visit to Holland in 1586. The *Terra Australis* entirely disappears (on Drake's authority) in the map attached to Hakluyt's second edition, 1598. Ortelius, however, continued to join Tierra del Fuego with the Terra Australis Incognita.

As they lounged over the bulwarks a sail appeared to the northward, and they made ready a pipe of wine to have a merry night with the new-comers. As the stranger anchored they beat her a welcome of their drum, and then watched her boat come alongside. In a moment all was in confusion. A rough old salt was laying about him with his fists, shouting in broken Spanish, "Down, dog, down!" and the astounded Spaniards were soon tight under hatches. It was Tom Moone at his old work. Hither the *Golden Hind* had been piloted by a friendly Indian in its search for provisions and loot. The little settlement was quickly plundered of all it had worth taking, and Drake's mariners, who for months had been living on salted penguin, and many of whom were suffering from wounds received in an encounter with the islanders of Mocha, were revelling in all the dainties of the Chilian paradise. For three days the mysterious ship, which seemed to have dropped from the skies, lay in the harbour collecting provisions, and then, laden with victuals, it sailed away northward with its prize.

Drake's great anxiety now was to rendezvous his scattered fleet for the sack of Lima and Panama, and assured that Wynter must be ahead he fully expected to find him in 30° north latitude, the point agreed on. After an ineffectual attempt to water at Coquimbo, where he found the Spaniards in arms, he discovered a natural harbour a little to the north of it which suited his purpose. In a month his preparations were complete. The men were thoroughly refreshed; a pinnace had been set up; the *Golden Hind* refitted from stem to stern, and under the guidance of the pilot of the *Grand Captain* he set out to realise the dream of his life. Every one, except perhaps poor John Doughty, was in the highest spirits. The return of health and the glorious climate made them reckless of the dangers of their single-handed attempt. Still they trusted to find the *Elizabeth*, and as they searched the coast for water with the pinnace they never lost hope of hearing of her. Fresh plunder constantly compensated for their continued disappointment. At one point on the coast of Tarapaca they found a Spaniard asleep with thirteen bars of silver beside him. They apologised profusely for disturbing his nap, and politely insisted on making amends by relieving him of his burden. Farther on they met another driving a train of guanacoes laden with some eight hundred pounds of silver, and expressing themselves shocked to see a gentleman turned carrier, they took his place; but somehow, as they afterwards said, they lost the way to his house and found themselves suddenly just where they had left the pinnace. So they romped along that peaceful coast, startling its luxurious slumbers with shouts of reckless laughter till they came to Arica, the frontier-town of Peru and the point where the fabulous wealth of the Potosi mines was embarked for Panama. It was a place important enough to have tempted the *Elizabeth* from her tryst. But not only was no trace of her to be found, but so hot was the alarm in front of Drake that two small treasure-barks were all there was in the harbour to plunder, and the town was in arms. A few hours ago a galleon had escaped northward, laden with eight hundred bars of silver, all belonging to the King of Spain, and fuming to so narrowly miss his revenge, Drake at once resolved to give chase. Without further care for his consort or any attempt on the town he hurried on with his pinnace and the Valparaiso prize, till at Chuli, the port of Arequipa, they saw the chase at anchor. Her capture was without a blow, for not a man was found aboard her —nor a bar of silver either. Two hours ago the whole of it had been heaved overboard to save it from Drake's hands, and in a fury of disappointment he at once set both the slow-sailing prizes adrift out into the ocean. For he was resolved by a dash on Lima to outstrip his notoriety at all costs, and so once more the *Golden Hind* and its pinnace spread their wings northward alone.

It was on February 15th that, in the dead of night, they quietly entered Callao de Lima. The harbour was full of shipping, and the pilot whom Drake had seized from a vessel outside was made to take him right in among them. A ship from Panama was entering at the same time, and as they anchored side by side, a custom-house boat at once put off and hailed them. Not content to wait till the morning, a sleepy officer boarded the *Golden Hind*, and before he knew where he was he tumbled right on the top of a big gun. Frightened to death, he was over the side again in a moment, and his boat dashed away crying the alarm. The ship of Panama cut her cables, and Drake slipped into the pinnace to take her; but as she showed fight he left her for the present and turned to ransack the defenceless shipping that lay around him. From ship to ship he went, but not an ounce of treasure could he find. It was all ashore except a vast quantity which had recently been shipped for Panama in a large vessel called *Our*

Lady of the Conception, and nicknamed the *Spitfire*.[1] That was enough for him. He returned to the *Golden Hind*, left his anchorage, and as he drifted out in the calm which had fallen, he captured the ship of Panama. But then ensued a delay both exasperating and dangerous. For three days there was not a breath of wind, and the Viceroy of Peru, marching down from Lima with two thousand troops, sent out four vessels to capture or burn the rover as he lay becalmed. All was in vain. Ere they found heart to close with the terrible stranger the breeze sprang up and away he went in hot pursuit of the treasure-ship. It had fourteen days' start of him, but he did not despair, and while the Viceroy was solemnly casting guns to arm vessels to pursue him, Drake was ransacking ship after ship for treasure and news of the chase. She had stopped at Truxillo to load more bullion, and each prize told him he was overhauling her. At Paita he learnt she had sailed but two days before. The scent was now hot indeed. Exasperated to miss his prey so narrowly, the Admiral promised a golden chain to the man who first sighted her, and swore she should be his. though he tore her from her moorings at Panama itself. Across the line they raced and still no sight of her, till on March 1st off Cape San Francisco young John Drake, his page and nephew, claimed the reward. Fearful of alarming his quarry, Drake at once ordered casks to be trailed astern, and so managed to keep hull down till nightfall. Then the *Golden Hind* was slipped, and in one bound rushed alongside her prey. A single shot brought

[1] See deposition of her master in *S. P. Spain*, 1580, bundle xviii, 9, b.

her to reason, and then side by side the two ships ran westwards for three days into the silent wastes of the Pacific. For three days more they lay together, and when they parted there were added to Drake's treasure thirteen chests of pieces of eight, eighty pounds weight of gold, jewels untold, and the *Golden Hind* was literally ballasted with silver.

So huge was the booty that the only thought was home. To attempt Panama single-handed would in any case have been madness, and Drake resolved to return, but not by the way he came. The great discoveries he had already made did not satisfy his greed for renown. He had swept one whole continent from the globe; by his survey of the coast of Chili he had for the first time determined the shape of another; and now he was minded to settle for ever the question of the North-West passage. From the Atlantic his rivals were seeking the fabulous Strait of Anian, and by that channel, if it existed, he determined to find his way home.

His daring resolve completely outwitted the Spaniards. The Viceroy of Peru sent his most brilliant officer, Don Pedro Sarmiente de Gamboa, in pursuit. He sought the rover towards Panama, but he was not there. Still ignorant that it was not the only passage between the two oceans, he turned to bar the way at the Straits of Magellan, and Drake was not there. But far away, in his palace at Mexico, Don Martin Enriquez, the perjured Viceroy who eleven years ago had broken his word at Vera Cruz, had news in plenty. Mocking greetings from his unknown enemy disturbed his ease, and he had to read news from the Nicaraguan coast that sorted ill with a quiet siesta. There a corsair, the like of whom no man had seen, had been at work. His prisoners had found him surrounded by a council of the younger sons of the first men in England, who always approached him hat in hand and stood in his presence. He dined in state to the sound of violins, and his crew, whose discipline filled the Spaniards with amazement, adored him. He was a martinet, and took no man's advice, but he heard all alike and had no favourite. He had artificers of every kind, and at the Isle of Caño had just careened and refitted his ship, God and His saints only knew for what fresh depredations. He had cartographers, who were making charts of the coast as he went, so that whole fleets might follow in his track. And as for catching him, so well armed and so fast was his ship that that was out of the question. The whole coast of New Spain was in a fever of alarm, for they knew it was the same Drake, the cousin of Aquinez, who, five years ago, had raided Nombre de Dios. The Bishop of Guatemala began melting his chimes into guns, ships were fitted out and troops moved up and down. In a month they expected to be ready to take the sea, but in a week Drake had done his work. Swooping on the port of Guatulco, he had found the court sitting, carried off all the judges bodily to his ship, and then made them send an order for every man to leave the town. This done, he revictualled at his ease from the Spanish storehouses, and next day he was away once more. He had less idea of staying than ever; for, lurking off the coast of Nicaragua, on the track of the China trade, he had made a capture of greater value than all his treasure. It was a vessel on which were sailing two China pilots, and now snug in the cabin of Spain's arch-enemy were the whole of the secret charts by which was conducted the rich Spanish trade across the Pacific.

For Spain it was a disaster of which no man could see the end, and, hugging his inestimable treasures, Drake sped northward to find his way back into the Atlantic. By the first week in June he had reached close-hauled on the north-east trade as high as the latitude of Cape Mendocino; but here he was suddenly caught in a storm of extraordinary severity. His rigging was frozen, his crew were half paralysed. Still he struggled on, firing his men with his own hot courage. In two days more he reached the latitude of Vancouver, and there he gave up the struggle. The land still trended westward, the weather grew more and more severe, and he made up his mind that if the passage existed it was impracticable. So the great resolve was taken, and running south to find a port to prepare the *Golden Hind* for her tremendous effort, he put into a natural harbour near San Francisco, where the cliffs were white like those at home, and the soil was teeming with gold. As fort and dockyard rose by their lonely shores the Indians gathered in wonder and would have worshipped the strangers as beings from a better land. The horrified Puritans protested as kindly as they might, and when persuaded Drake was human, the simple savages crowned him in his mistress's name king of New Albion. So at least the old navigators understood the strange ceremonies with which the month of their stay was occupied; and the loud lamentations of their friends when they departed filled their imaginations with visions of an empire of Englishmen hardly less grand than the great reality.

It was on July 25th that, with a boldness we can hardly realise, the course was laid direct for the Moluccas. Their instruments for finding latitude were far from perfect; longitude it was practically impossible for them to determine at all; their logs were so distrusted that as a rule they preferred to guess the runs; and the variation of the compass was ascertained with childish crudeness. Yet Drake did not even condescend to follow the beaten trade-track of the Spaniards along the ninth parallel. But straight across the Pacific, from where he was to where he wished to be, he pushed his way as it were by inspiration. For sixty-eight days they had no sight of land. By the end of September they found themselves close to the equator, and turning to the northward to avoid the counter-current, on the last day of the month they ran in amongst the Carolines.

The rest is long to tell; how, getting clear of the pilfering natives, Drake made the Philippines, and coasting along them ran from the southern point of Mindanao through the Talautse group and past Togolando to the Moluccas; how at Ternate he made an exclusive commercial treaty with the king which, for a century afterwards, was the sheet-anchor of our diplomatists in their quarrels with the Dutch and Portuguese about the East Indian trade; how he careened again at an island near the Greyhound Strait, and then, after trying to beat northward into the Macassar channel, turned back to pass southward, and was at once entangled in the reef-encumbered seas that wash the eastern coasts of Celebes; and how, after escaping a thousand dangers in the first days of the year 1580, as they were sailing along the south of Peling Island with a fine topsail breeze, they ran full tilt on a reef. There for twenty hours they lay at the mercy of God. All around was deep sea, where no hold could be got for warping. Every shift was tried, but not an inch would the treasure-laden vessel stir, and death only grew more real before them. Hopeless and exhausted, they desisted from their efforts, and in solemn preparation for the end, took the Sacrament together. Then in the good old Puritan fashion, to aid the Lord, Drake made jettison of guns and spices worth their weight in silver, till lo! in the midst of their pious labour the wind changed, and, like the breath of the Saviour in answer to their prayers, gently slid them from the rock. It was the gravest danger of all their voyage, and for nearly two months more, as they groped their way about the Floris Sea and struggled with baffling gales, they hourly expected its recurrence. But every peril was overcome at last, and in March they were well clear of the Archipelago, and with thankful hearts refitting, cleaning, and victualling in a southern port of Java. So the great exploit was accomplished, and the prayer uttered so devoutly six years ago upon the giant tree in Darien was more than fulfilled. God had given His supplicant life and leave to sail the South Sea in an English ship, and he had sailed it from side to side. Its secret was England's at last; and, laden with its wealth, in two months more the triumphant explorer was ploughing his homeward way towards the Cape of Good Hope.

CHAPTER VII

THE GREAT MISTAKE

At home, for more than a year, nothing had been heard of Drake beyond a rumour that he had been taken by the Spaniards and hanged. That the Doughty plot had failed, and that the Queen's chief pirate had entered the South Sea, was known to the Government. Wynter had brought the news in the summer of 1579, together with a report of the chaos of storm into which the *Golden Hind* had disappeared. Burleigh and the party with which he was acting were consoling themselves with the reflection that their bugbear must certainly have gone to the bottom, when one day towards the end of August, Mendoza, the Spanish Ambassador, sprang upon them the astounding news which the Viceroy of Mexico had just had the pleasure of sending home. The peace-party were aghast. Except the vague report of Drake's execution, there seemed nothing between them and war, and Walsingham could rub his hands over the success of his scheme. Philip, however, was as slow as Elizabeth was irresolute. Moreover his hands were full of his preparations to seize the inheritance of the dying King of Portugal, and Mendoza was instructed to accept the Queen's excuse. She had protested that Drake was a private adventurer, and that she had nothing to object to his alleged execution, and that if he ever came home he would be severely dealt with. So all the warlike ambassador could do was to sit down and watch like a lynx for the pirate's reappearance.

Such was the welcome that awaited the *Golden Hind*, as one day about Michaelmas 1580, worm-eaten and weed-clogged, she laboured into Plymouth Sound.[1] Drake, no doubt, was ready, as he always was, to be disowned, and if necessary to disappear; and no sooner had his anchor plunged triumphantly into the well-known waters than his friends warned him of his peril. With characteristic decision he immediately warped out his priceless burden behind St. Nicholas Island, and making the plague which had been raging in Plymouth his excuse, refused to land. There he lay defiantly while his messenger sped to Court, and amidst the wonder of his friends and the caresses of his wife he resolutely awaited the end.

In a week came a summons to Court. Drake obeyed at once, but he did not go alone. He had received letters of advice from his great friends, and in his company went a train of pack-horses laden with the cream of his plunder. It was oil to assuage the storm which Mendoza was brewing in the Council. Burleigh and the more earnest of the peace-party refused his gifts with scorn, but still Drake had no cause to lose heart. He reached London in a moment so happy that it would almost seem again as though some fairy godmother had

[1] So secret was Drake's arrival kept that every authority assigns to it a different date, varying from September 16th to November 3rd. Probably he arrived September 26th, and landed October 3rd or 4th.

waved her wand to save him. The Council had recently learnt for certain that Philip had seized Portugal almost without a blow, and that Don Antonio, the pretender to the throne, had been hunted from the realm like a wild beast. To the fabulous power and wealth he already possessed, Philip at a stroke had added the vast possessions of Portugal. The whole world seemed stretched at his feet. The great Armada in Cadiz was still unused, and free to strike no one knew where. Fearing the worst, the Council with feverish haste were pushing forward military preparations, when the very week that Drake came home news reached London that a Spanish force had landed in Ireland. England's hour seemed come at last. Half the Council were in despair, and ready to have peace on any terms, when suddenly the tone of English diplomacy changed to a ringing note of resistance. For the Queen's little pirate was whispering in her ear, as he walked beside her relating his adventures, that though the whole world were the King of Spain's garden, yet it was hers to pluck the fruit as she would.

Mendoza raged, but nothing could he get save an order to Edmund Tremayne at Plymouth to register the plunder and send it up to London. Tremayne was an old confidant of the Queen's; and on the heels of the order Drake followed with a private letter under her sign-manual telling the Devon justice to turn his back while her well-beloved subject secretly abstracted ten thousand pounds' worth of bullion for himself. It was a sum equal to about eighty thousand pounds of our money, and this was part of his reward.

Still the Queen would not openly recognise him.

Though she refused to see Mendoza while a single Spanish musketeer remained in Ireland, still in her usual way she attempted to allay his anger. An inquiry was held at which the whole of Drake's crew swore that not a single act of cruelty had been committed as the Spaniards alleged, and so much of the plunder as Tremayne chose to see was registered, sent up to the Tower, and a return given to the Ambassador. But each day he grew more furious. Drake came back to town, bringing up the *Golden Hind* for all the world to stare at. He was the lion of the hour, and every day had long interviews with the Queen, so charmed was she with her little pirate and what he had brought back. In vain Mendoza and his friends tried to frighten the London merchants with the prospect of war, or an embargo at least. The splendour of Drake's achievement, and the richness of his presents, turned every one's head. Rumour swelled the amount of the plunder to a fabulous extravagance. His very crew seemed princes in their splendour. Two years ago, beside Doughty's grave, they had agreed to stand to their Captain's courtesy, and so well had he treated them that the people ran out of their houses as they swaggered by, to wonder at their wealth. So, in spite of all Mendoza could threaten, louder and louder grew the exultation as the news of the triumphant exploit spread through the realm and waked in the farthest hamlet the old fierce spirit of the Hundred Years' War.

Still Mendoza, with the support he had in Council, was dangerous, and Leicester and his fellow-shareholders tried to bribe him. The attempt only made his complaints the more indignant, and as month after month of intrigue went by, people were still uncertain whether, after all, Drake or his treasure would not have to be given up. But Elizabeth never doubted a moment; she and Walsingham had fresh work for him to do.

Terceira, one of the Azores, had refused to recognise the Spanish conquest, and Don Antonio, having escaped to Paris with his life and the Braganza jewels, was crying lustily for help to go and support his faithful subjects. Walsingham saw the opportunity. Drake and his contemporaries were lifting naval warfare to a new level, and the far-sighted Foreign Minister was the first to learn the lesson their exploits taught. Hitherto the navy had been the mere handmaid of the military, and had been used by belligerents for nothing more intelligent than raids upon each other's coasts. But now the wisest heads in England began to divine for it a higher function than the paltry cross-ravaging of the Middle Ages, and to see the tremendous weapon a powerful fleet would be in the hands of the power that first used it against an enemy's trade. It is Drake's chief claim to be called a great admiral, that he was the pioneer of this strategic revolution. Above all countries Spain was most exposed to such an attack, and he saw that the moment had come. Terceira was on the direct road of the fleets returning both from the East and West Indies, and Drake told the Queen if he might seize it in Antonio's name and act from it as a base, he could so utterly destroy the trade of Spain that she would be able to dictate her own terms of reconciliation. Elizabeth seized the bait. Whether or not she grasped the whole significance of Drake's proposal, a privateer war in a pretender's name was irresistible to her tortuous statecraft. By March it was already whispered that Mr. Drake was going on a second voyage, well furnished and countenanced. Walsingham was busy working out alternative projects, and in April the Queen threw off all disguise. Drake's worm-eaten ship had been hauled ashore at Deptford, and on the 4th of the month the Queen went down, and after a banquet on board publicly knighted the "master-thief of the unknown world." The *Golden Hind* was ordered to be preserved for ever, as a worthy rival of Magellan's *Victoria*, and all the world knew at last that the "King of Spain's long legs" were in sore danger.

In July the expedition was ready to sail. Hawkins and Drake had arranged it together. Courtiers and merchants had freely subscribed the cost, and every adventurous spirit had flocked to their flag. Norreys, Bingham, Carleill, and all the rest were there, and Fenton and Yorke, Frobisher's favourite lieutenants, were each in command of ships. Don Antonio had come over, and after a secret interview with the Queen had pawned his jewels and gone down to Plymouth. Nothing was wanting but the order to sail; but day after day went by and nothing but scolding came from the Queen. As usual with her on the eve of doing something heroic, she was losing heart and fretting herself into a state of nervous irritability. She was angry because Drake had spent £2000 more than the estimate: she could not get a definite answer from the French king that the contingent he had promised was going to sail; and she was fast persuading herself that by a renewal of her flirtation with the Duc d'Alençon she could frighten Spain into condoning Drake's offences. As the days went by in inaction Drake lost his temper too. No doubt he was not less masterful since his brilliant reception, and his splendid self-reliance was more than ever impatient of restraint. Pardonably jealous of his sudden predominance, the Irish officers and Frobisher's men found unendurable his habit of listening to every one's advice and always taking his own. Yorke was even suspected of treachery, and by the middle of August, when sailing orders were hourly expected, so acute had the friction grown that they all found it necessary to resign their commissions.

Drake and Hawkins had triumphed. The whole affair was now in the hands of their own set; but still no letter came from the King of France, and in a week, afraid to act and afraid to draw back, Elizabeth had hit on one of her disastrous compromises. She resolved to send a little squadron under Norreys and young William Hawkins, and to pay off the rest of the fleet. Don Antonio refused such pitiful patronage, and after vainly trying to recover his jewels, retired to France in a rage. William Boroughs, of Muscovy fame, came down with a commission to break up the expedition, and so ended one of the most brilliant projects that a far-sighted strategy ever hatched.[1]

Drake's deep disappointment is easy to picture; but the Queen would not hear of either his or Hawkins's leaving the country. Till she had pushed France into the fire she did not mean to provoke Spain further, and till Spain was occupied with a French war she wanted her best men at her side. So in organisation and routine work Drake had plenty to do for his wayward mistress. He was chosen Mayor of Plymouth too, and was soon busy with projects for its improvement and defence.

[1] Digge's *Compleat Ambassador*, 379 *et seq.*

Nor was this all. Leicester, bent on saving something from the wreck of the abortive venture in which he had been deeply engaged, resolved to organise from its ruins an expedition to carry out one of Walsingham's alternative schemes. It was for a trading and exploring voyage into the East Indies to widen the opening which Drake had begun. Again an attempt was made to combine the two sets, and the command was divided between Fenton and William Hawkins. Drake, like most other people, knew how incapable Fenton was; but at Leicester's entreaty he loyally did his best both with money and advice to make the thing a success, and Fenton handsomely acknowledged his generosity. So the expedition sailed next year with all the prestige of the great navigator's support; but Fenton's stupidity, and his jealousy of his brilliant young lieutenant and the rest of Drake's men, proved invincible, and the affair ended in complete disaster.

Thus it was that as well as he could, surrounded by treachery and jealousy, Drake tried to carry on his life's work. When the Queen had finally refused to give him up, Philip too had recognised the feud. He was keen enough to see in the Puritan corsair an arch-enemy, and diplomacy failing, he stealthily drew his favourite weapon. Knowing well that Drake had enemies, and above all that Doughty's brother was living for revenge, he offered a sum of more than forty thousand pounds of our money to any one who would assassinate or kidnap the renowned corsair. Spanish owners who had suffered from Drake's depredations, with faith unshaken in his clemency to private citizens, were sending over an agent to make terms with him. This man was charged with the mission, and he at once approached John Doughty, who had naturally found it impossible to get any redress for his brother's death. For obvious reasons Drake's opponents in the Council dared not call the execution in question. They had even to assent to its procedure being used as a precedent in Fenton's official instructions. Doughty, therefore, eagerly embraced his only chance of revenge. The matter was settled; and of course Doughty began boasting of what he was going to do, just as before he had boasted of his skill in the Black Art. Drake soon heard of his folly, obtained his arrest from the Council, and a spy of the Spanish agent's confessed the whole crime on the rack. So the tragedy ended. Doughty remained in prison, and save for one despairing cry for his release, so far as we know, was never heard of again.[1]

But so far from its being the end of Philip's projects, he was more than ever bent on revenge for all the stripes he had received. While the year 1583 wore away in ominous quiet, while Drake was mourning the loss of his wife, and as a member of the Navy Commission was arranging a peace-footing for the fleet from which it could be rapidly mobilised, the arch-schemer of the Escurial was piling up his great French plot to crush his island enemy. In November it all came to light, and from Scotland, France, and Spain, a simultaneous invasion was threatening upon a country scored with treason. Then at last the Queen awoke, and while the fleet in three divisions watched for the coming blows, the nation armed. Mendoza was expelled, the seaports fortified, the border

[1] *S. P. Dom. Eliz.* cliii. *f.* 50, and clxiii. *Hatfield MSS. Cal.* vol. ii. 515.

strengthened, and Catholics arrested wholesale. The immediate danger passed, but the effect of the shock remained. Burleigh was now all for action, and ere the year was out, hands had been joined with the Prince of Orange. Drake as member for Bossiney was voting supplies and death to traitors at Westminster; and in his pocket were orders that were destined to mark the point where the tide of Spanish Empire turned.

CHAPTER VIII

THE DRAGON LOOSED

WARNED by his disappointment of three years ago, Drake threw himself with all his vigour into the work of organising his force. His parliamentary duties still engaged his attention, and in February, 1585, he was married to Elizabeth Sydenham, the young heiress of a knightly and warlike house in Somersetshire. But nothing was allowed to interfere with his mission. As early as November in the past year Walsingham had summoned his son-in-law, Captain Christopher Carleill, from Ireland to command the troops, and in the early spring things were so far advanced that Philip took alarm. His imagination multiplied the havoc which Drake had made with one ship by the numbers of the gathering squadron, and he trembled for his Indies. His alarm was premature. Once more as Elizabeth felt the hot breath of war upon her cheek she shrank from its horrors like the very woman she was. Drake's work was stopped, and he was left to fret as the weary diplomatic dance began again. But beneath it all the mill of Philip's purpose went grinding on relentlessly. He used the lull only to invite a large fleet of English cornships to the relief of his famine-stricken provinces, and then as they lay unsuspecting in his ports he seized them every one. Never once was the growing Armada out of his mind. This atrocious outrage was but to feed his monster, and swift and sharp was the retribution it earned. It was in the last days of May, and ere June was out far and near the seas were swarming with English privateers, and El Draque was unchained.

Fortified with letters of marque to release the embargoed vessels, he hoisted his flag at Plymouth on the *Elizabeth Bonaventura*, and there, by the end of July, "in all jollity and with all help and furtherance himself could wish," a formidable fleet gathered round him. Frobisher was his vice-admiral, Francis Knollys his rear-admiral, and Thomas Fenner his flag-captain. Christopher Carleill was there too as lieutenant-general, with a full staff and ten companies under him. No such privateering squadron had ever been seen before. It consisted of two battle-ships and eighteen cruisers, with their complement of store-ships and pinnaces; it was manned with a force of soldiers and sailors to the number of two thousand three hundred, and it is not surprising that constant difficulties delayed its departure. Yet delay was dangerous in the extreme. The Spanish party had again taken heart, and were whispering caution in the Queen's ear. Even Burleigh grew nervous that she would repent; but at last he got sailing-orders sent off, and, with a sigh of relief, entered in his diary that Drake had gone. To his horror came back a letter from the Admiral still dated from Plymouth, instead of from Finisterre, as he had hoped, and he sent down a warning to urge the immediate departure of the fleet. August wore away, and the equipment was still incomplete, when Drake, who was now in constant dread of a countermand, was alarmed by Sir Philip Sydney's suddenly appearing at Plymouth and announcing his intention of accompanying the expedition. Haunted still by the memory of the tragedy he could never forget, and determined to have no more to do with courtiers and amateur soldiers, he secretly sent off a courier to betray the truant's escapade to the Court. He must even be suspected, in his desperation, of having set men in wait to intercept and destroy any orders that were not to his liking.[1] The precaution was unnecessary. Sydney was peremptorily stopped, and ere any letter came to stay Drake too, the wind had shifted northerly, and all unready as he was he cleared for Finisterre.

There he arrived on September 26th. He was clear away, but that was all. He was short both of water and victuals. There had not even been time to distribute the stores he had, or to issue his general orders to the fleet. He smelt foul weather, too; and, determined to complete somewhere what he had left undone at Plymouth, he boldly ran in under the lee of the Bayona Islands in Vigo Bay. The old Queen's officers were aghast. Entirely dominated by the prestige of Spain, they believed that nothing could be done against her except by surprise, and they trembled to see their Admiral thus recklessly fling his cards upon the table. But he knew what he was doing. As with sagacious bravado he had sprung ashore at Santa Marta, and had mocked the Spanish fleet in Cartagena harbour,

[1] *Hist. MSS. Com. Rep.* XII. iv. 178, 180. The fantastic complexion which Fulke Greville put on this incident can only be excused by his infatuation for his friend. His story not only contradicts itself, but is at issue with almost every State Paper which touches the question.

so now before he struck he exulted that his unfleshed host should hear him shout *en garde!* to the King of Spain; that they should listen while he cried that England cared not for spying traitors, for she had nothing to conceal, that her fleets meant to sail when and where they would, and Philip might do his worst. It was a stroke of that divine instinct which marks out a hero from among able captains;—the magic touch of a great leader of men, under which the dead fabric of an army springs into life and feels every fibre tingling with the strong purpose of its heart.

Two leagues from the town of Bayona the fleet anchored; and resolved at once to display his whole strength, and exercise his men in their duties, Drake ordered out his pinnaces and boats for a reconnaissance in force. His boldness bore immediate fruit. The Governor sent off to treat, and by nightfall it was arranged that troops should land, and in the morning be allowed to water and collect what victuals they could. But at midnight the threatened storm rolled up. The troops were hurriedly re-embarked; and barely in time to escape disaster, the flotilla regained the ships. For three days the gale continued, threatening the whole fleet with destruction till it was got safely up above Vigo. There the whole of the boats in which the panic-stricken inhabitants had embarked their property were captured, and though by this time the Governor of Bayona had arrived with a considerable force, he was compelled to permit Drake to carry out his purpose in peace.

By October 8th he was out in the Bayona road again, waiting for a wind to waft him on his way, and it was reported at the Spanish Court that he had gone towards the Indies. The consternation was universal. The Marquis of Santa Cruz, High Admiral of Spain and the most renowned naval officer in Europe, declared that not only the African islands, but the whole Pacific coast, the Spanish Main, and the West Indies, were at the corsair's mercy, and told his master that a fleet of forty sail must be instantly equipped for the pursuit. But though for another fortnight Drake rode defiantly at the Bayona anchorage, not a limb of Philip's inert machinery could be moved against him; and while the chivalry of Spain chafed under their sovereign's deliberation, the second blow was struck.

Madeira was passed by and the Canaries spared; for Palma, which Drake intended should revictual him, showed so bold a front that he would not waste time in trying to reduce it. It was on another point that his implacable glance was fixed.

Five years ago at Santiago, the chief town of the Cape Verde Islands, young William Hawkins, a personal adherent of Drake's, had been made the victim of some such treachery as his father and captain had suffered together at Vera Cruz. From that hour it was doomed. In the middle of November the fleet arrived in the road, and the troops landed. Threatened by Carleill from the heights above the valley where it lies, and from the sea by Drake, without a blow the town was abandoned to its fate. For ten days the island was scoured for plunder and provisions, and ere the month was out the anchorage was desolate and Santiago a heap of ashes.

Drake's vengeance was complete, and exulting like Gideon in the devastation that marked his course, he led his ships across the Atlantic. Is there a moment in history more tragic than that? For the first time since the ages began, a hostile fleet was passing the ocean—the pioneer of how many more that have gone and are yet to go—the forerunner of how much glory and shame and misery! What wonder if the curse of God seemed upon it? Hardly had it lost sight of land when it was stricken with sickness. In a few days some three hundred men were dead, and numbers of others prostrate and useless; but in unshaken faith and with reverent wonder at the inscrutable will of Heaven, Drake never flinched or paused. His only thought was how to check the evil. At Dominica he got fresh provisions from the natives, and refreshed his sick with a few days on shore. At St. Christopher he again halted to spend Christmas and elaborate the details of his next move. The point where Philip was now to feel the weight of his arm was the fair city of St. Domingo in Hispaniola. It was by far the most serious operation Drake had yet undertaken. Hitherto his exploits had been against places that were little more than struggling settlements, but St. Domingo was indeed a city, stone-built and walled and flanked with formidable batteries. It was held by a powerful garrison, as Drake learned from a captured frigate, and a naval force had been concentrated in the harbour for its defence. As the oldest town in the Indies, its renown had hitherto secured it from attack, and in Spain it was held the queen-city of the colonial empire. The moral effect of its capture would be profound, and besides, from Virginia the governor of Raleigh's new colony had sent home a fabulous report of its wealth. Drake was fully alive to the gravity of the task before him. His dispositions had never been so elaborate, and they evince at least a touch of that military genius which the strategists of the next century denied him. While the sick were recruiting he sent forward a squadron to reconnoitre, and, if possible, to open communications with the Maroons who infested the hills. For three days the garrison was thus exhausted with constant alarms, and then on January 1st, 1586, the whole fleet appeared in the bay.

Night fell, and as darkness closed the eyes of the harassed garrison, with the fleet all was activity. In boats and pinnaces the troops were being rapidly embarked, and soon Drake in person was piloting the flotilla for the surf-beaten shore. At a point within the bay, but some ten miles from the town, a practicable landing-place had been found. Watch-houses overlooked it, but watchmen there were none. Drake had got touch with the Maroons. By his directions a party of them had stolen down from the hills, and as the sentries came out from the city in the evening, swiftly and silently they had been every one despatched. Thus unseen and unmolested, the troops were successfully landed, and then with pious and cheery farewells to Carleill, Drake returned to the fleet to prepare the ground for the surprise.[1]

In the morning he anchored in the road, ran out his

[1] "Advertisement concerning Sir Fr. Drake," *S. P. Dom. Eliz.* clxxxiii. 27, and cf. *Ib.* f. 47. Cates, who wrote the only connected account of this expedition that has come down to us, was a company officer who never fully appreciated the significance of the movements in which he was engaged. Though a warm admirer of Drake, he practically ignores on every occasion the naval part of the operations.

guns, and proceeded to threaten a landing at a point close to that side of the town upon which Carleill was stealthily approaching in two parallel columns. As the Spaniards saw the fleet preparing the advance of the boats and pinnaces, the whole of the horse and a large force of foot marched out of the town to oppose the threatened attack, and took up a position fronting the sea, with their left resting on the town and the other flank exposed in the line of Carleill's advance. It was exactly what had been foreseen, and ere the Spaniards had discovered that the movement from the fleet was merely a feint, the horse which were covering their exposed flank were flying before Carleill's musketeers. The surprise was complete. Taken in flank by Carleill, and threatened in the rear by his second column under Powell, the chief of the staff, the infantry could make no real resistance; and so rapidly was the English advance pushed home, that the struggling mass of friend and foe entered pell-mell through the open gates of the town. For an hour alarms of drum and trumpet mingling confusedly with the sounds of street-fighting reached the listening fleet, as the two columns forced their way to meet upon the Plaza. But how they fared none could tell, till on a tower a white staff suddenly appeared, and in another moment the cross of St. George fluttered gaily out upon the breeze. With a roar of triumph the ship's guns saluted the signal of victory. The town was won.

Though the garrison fled panic-stricken across the river on the far side of the city, and the citadel was evacuated in the night, the place was far too large to be occupied by the force at Drake's command. Following therefore the same tactics that had been successful at Nombre de Dios, he ordered the troops to intrench themselves in the Plaza, and to occupy the principal batteries. In this way he held the city for a month. The plunder was disappointing. The city was already a hundred years old, and its day was done; for the reckless native policy of the colonists had almost ruined the island. It remained but to treat for a ransom The Governor at once declared himself unable to meet the extravagant demands of the English Admiral, and in order to bring him to terms Drake began to burn the town piecemeal. But so well was it built, that little harm could be done, and every day his impatience increased. Once in the course of the negotiation he sent a boy with a flag of truce to the Spanish camp. A Spaniard, meeting the lad, so ill-treated him that he could barely crawl back to die at the Admiral's feet. Then all the fury of Drake's nature burst forth. Two friars who were among the prisoners were immediately sent ashore and hanged by the provost-marshal on the scene of the crime. Another was despatched to the Spanish camp to declare that two more would be executed every morning until the offender was brought down and hanged on the spot by his own authorities. In hasty alarm the demand was complied with, and then the international dinners and the negotiations went on more smoothly. Convinced at last of the poverty of the colony, Drake accepted a ransom of twenty-five thousand ducats. The sum, which is equal to about fifty thousand pounds of our money, though little enough to satisfy the shareholders, was very serious for the enemy. For besides this loss the town had been stripped of everything worth carrying away by the troops and seamen. Two hundred and forty guns were taken on board the English ships, and not only were they thoroughly refurnished from the Spanish stores, but for a month the whole expedition had lived in free quarters at the enemy's expense. The entire fleet which lay in the harbour fell into Drake's hands, and with the exception of four of the finest galleons was given to the flames. Besides the vessels which the Spaniards themselves had scuttled, two galleys with their tenders, fifteen frigates, and a galleon were thus destroyed, and hundreds of galley-slaves set free.

"It was such a cooling to King Philip," said one in Europe as the news leaked out, "as never happened to him since he was King of Spain." But as yet Drake was far from done. In the middle of February, with his force recruited by the English prisoners he had freed, and with a troop of attendant prizes laden with his spoil, in undiminished strength he appeared before Cartagena. No city in America was more difficult of approach, but the memories of the old hard days were still green, when storm-beaten, drenched, and chilled, without food or shelter, he had ridden in the harbour day after day in despite of all the Spaniards could do, and he knew it all like a pilot. The city was built close to the shore fronting west, and directly from its southern face an inlet of the sea stretched many leagues southward along the coast, forming a large lagoon. The long spit of land which separated this sheet of water from the sea was pierced by two natural channels. At the far end was the dangerous *Bocca Chica*, and some three miles from the city was a larger entrance known as the *Bocca Grande*.

Between this entrance and the town a tongue of land ran out at right angles from the spit to the opposite shore, forming an inner harbour and barring all approach to the city from the outer part of the lagoon, except by a narrow channel which lay under the guns of a powerful fort on the mainland. On its northern and eastern faces the city was encircled by a broad creek, which ran round it from the inner harbour to the sea in such a way as to form a wide natural moat, rendering the city unapproachable from the mainland except by a bridge. This bridge was also commanded by the harbour fort, nor were land operations possible at any other point except from that part of the spit which lay between the city and the *Bocca Grande*. So finely, however, did this narrow down before the city could be reached, that between the inner harbour and the sea it was but fifty paces wide, and here the Spaniards had had time to prepare defences that looked impregnable. From shore to shore a formidable entrenchment completely barred the way, and not only was its front so staked and encumbered as to render a night attack impossible, but its approaches were swept by the guns and small-arms of a great galleasse and two galleys which lay in the inner harbour.

To a man so tender as Drake ever was for the lives of his men and the safety of his ships, to attack such a place might well have appeared hopeless; but the originality of the amphibious corsair at once descried a hole which had escaped all the science of the Spanish martialists. Instead of entering by the *Bocca Grande*, with consummate skill and daring he piloted the whole fleet through the dangerous channel at the extreme end of the lagoon. The only impression which so hazardous a movement could create in the minds of the Spaniards was that he was about to repeat his St. Domingo operations, and land his troops there to attack from the mainland. Such an impression must have been confirmed as, moving up the lagoon, he anchored opposite the *Bocca Grande* and threatened the harbour fort with his boats; but Drake's project was far different. Instead of being landed on the mainland, Carleill with eight companies was quietly slipped ashore in the *Bocca Grande*, with instructions to make his way diagonally through the woods that covered the spit till he reached the seashore, and then, instead of advancing on the front of the intrenchments, to wade along through the wash of the surf till he was within striking distance of the Spanish position. Meanwhile Frobisher advanced with the flotilla against the harbour fort, and as soon as Carleill was heard in contact with the enemy's pickets he opened fire. The boat-attack was repulsed—indeed, it may only have been intended as what soldiers then called "a hot alarm"—but Carleill was completely successful. By the march through the surf he had not only evaded the obstacles which the enemy had so carefully prepared, but he had been covered from the fire of the galleys in the harbour, and had never so much as entered the fire-area of the heavily armed intrenchment. After a desperate struggle at push of pike the position was carried by assault, and once more so hotly was the advantage pursued that in one rush the whole town was captured. The garrison fled across the bridge to the hills, and next day when Drake brought up the fleet to bear upon the fort that also was evacuated.

No success was ever better earned, and few more richly rewarded. Cartagena was the capital of the Spanish Main, and though much younger than St. Domingo it was far wealthier. It yielded rich loot for the men, and for his shareholders Drake after a long negotiation succeeded in exacting a ransom of a hundred and ten thousand ducats, besides what he got for an adjacent monastery. Though to all this plunder Drake could add the consolation that he had destroyed the galleys and shipping which crowded the port, and blown up the harbour fort which the Spaniards had forgotten to include in the convention, he was still unsatisfied. Well knowing that by an advance up the Chagres river in his boats Panama lay at his mercy, he was resolved with its capture to crown the campaign; but as he lay in Cartagena the sickness, which had never really ceased, broke out again with new virulence, and made such havoc with his force that he had reluctantly to confess that Panama must wait. To capture it with the crippled means at his command was impossible, and the only question was whether Cartagena should be held till he could return with reinforcements. The soldiers declared themselves ready to undertake the task; but in a full council of war it was finally decided that no strategical advantage would be gained at all proportional to the risk that would be run in further weakening the fleet, and on the last day of March the signal to make sail home was flying from the *Elizabeth Bonaventura*. So severely, however, did they suffer from the weather and want of water that it was nearly two months before they reached the coast of Florida. Still Drake found time and energy to destroy and plunder the Spanish settlement at St. Augustine, and relieve Raleigh's exhausted colony in Virginia. With the remnants of the settlers on board he weighed for England, and on July 28th, 1586, he was writing from Plymouth to Lord Burleigh laconically reporting his return; and apologising for having missed the Plate fleet by only twelve hours' sail—"the reason best known to God"—he declared that he and his fleet were ready at once to strike again in any direction the Queen would be pleased to indicate.

CHAPTER IX

SINGEING THE KING OF SPAIN'S BEARD

"THERE is a very great gap opened," said Drake in his letter to Burleigh, "very little to the liking of the King of Spain." That, with the calm request for orders, was his comment on a feat which changed the destinies of Europe. At its fullest flood he had stemmed the tide of Spanish Empire. It was no less a thing than that.

A few months ago all Europe had been cowering in confused alarm before the shadow of a new Roman Empire. Ever since the first triumph of Luther the cause of Reformation had been steadily losing ground; on England and the Low Countries hung its only hope, and with the fall of Antwerp Europe saw itself on the eve of that "last great battle in the west" which must decide its fate for centuries. In despair of the result each trembling Power was trying to hide behind the other: each was thrusting its neighbour forward to break the coming blow; and Philip led the cheating till his hour should come. He was bent on crushing Elizabeth; and then with one foot on the ruins of her kingdom he meant to stamp down his rebellious Netherlands into the gloomy Catholicism in which his own dark soul was sunk. As the fruit of his splendid deliberation ripened he strove to cheat her into inactivity by a hope that peace might yet be purchased by the betrayal of the Netherlands. She bit the bait and spat it out, and bit again, and all the while squirted round her a cloud of falsehood as black as that with which Philip was covering his spring. Her wisest councillors were in despair at her folly, and confessed to each other that on Francis Drake hung the last hope of Europe.

Then in laughing gusts came over the Atlantic the rumours of his exploits, till the full gale they heralded swept over Europe, whirling into oblivion a hundred intrigues and bending the prestige of Spain like a reed. The limitless possibilities of the new-born naval warfare had been demonstrated, and the lesson startled Europe like a revelation. An unmeasured force was added to statecraft, and a new power had arisen. The effect was immediate. Men saw the fountain of Spanish trade at England's mercy; they knew how narrowly the Plate fleet had escaped, and a panic palsied Philip's finance. The Bank of Seville broke; that of Venice was in despair; and the King of Spain, pointed at as a bankrupt, failed to raise a loan of half a million ducats. Parma was appalled. With his brilliant capture of Antwerp he had seen himself on the brink of that great exploit with which he hoped to crown his career; and now, instead of a host armed at all points for the invasion of England, he saw around him a broken army it was impossible to supply. In Germany the Protestant princes raised their heads, and seeing dawn at last, began to shake off the lethargy into which despair had plunged them. England was wild with joy. Burleigh himself was almost startled from his caution, and cried out with half a shudder that Drake was a fearful man to the King of Spain.

But so tumultuously was the great epic now hurrying to its catastrophe that Drake could not be spared a moment from the scene. In the midst of the ovation with which he was received, the great Babington Plot was disclosed. It was known that Philip, by a combined operation from Lisbon and the Netherlands, had intended to invade England the moment he heard Elizabeth had been assassinated, and many believed he would persevere in spite of Babington's failure and Drake's triumphant return. The imprisonment of Mary Stuart for her complicity in the plot was followed by a threat of war from France, and no one could tell what Scotland would do. The fleet was mobilised to watch the narrow seas, and so great had been the anxiety of the Government while Drake was away that they had no idea of letting him cross the ocean again till the danger was over. He knew well enough that in attack lay England's best defence. His genius had discovered how a naval power should make war, and he was craving for leave to deal another stroke at Philip's trade. But so far from being permitted to repeat his blow, Elizabeth sought to pacify Philip by a brazen disavowal of his late exploits. This was his reward. Peerages and pensions the Faëry Queen kept for her carpet-knights. The fighting men had to rest content with plunder and renown; and Drake cheerfully accepted a position for which his loyalty was fully prepared and which only made him more conspicuously a factor in European politics.

To the consternation of the whole Catholic world,

Philip accepted Elizabeth's transparent excuse. Disgusted at Babington's failure, he was resolved that nothing should again tempt him from his own line of approach. His method was slow and laborious, but time alone was wanted to make success certain. So the smiles and the lying went on again; and while Philip and Parma, under cover of diplomacy, resumed the gigantic preparations which Drake had interrupted with a panic, Elizabeth turned once more to her little pirate. Since his return he had been condemned to the old futile tactics, and with Sir William Wynter had been keeping watch in the narrow seas for the invasion that was hourly expected. But as the autumn waned the immediate danger passed; the Channel squadrons were brought in to be overhauled, and Drake hurried over to the Low Countries on a secret mission.[1]

A joint expedition with the Dutch against the Spanish Indies had long been urged by the Queen's most far-seeing statesmen and soldiers. They knew it would give Spain a wound so deadly that she never could be the same again; and now the moment seemed arrived. Late in the autumn Drake crossed the seas that he knew so well from the hard days of his boyhood. Wherever he went, the great navigator was received like a prince, and although the States refused him assistance officially, he was authorised to treat privately with the various cities. Men who knew him had no doubt of his success, and all would have gone well could Elizabeth have been for a moment sincere. But the distrust which Leicester's fatuous government had engendered, almost from the moment he had been sent in answer to the

[1] *S. P. Dom. Eliz.* cxciii. 49.

rebels' prayer for a leader, was only deepened by his withdrawal from the scene of his failure. He returned to England in November, and Drake with hopes still high followed him to organise the English contingent for the new enterprise. But whatever the Dutch cities may have intended, all hope of co-operation was at an end when, in January, Deventer and the fort of Zutphen were betrayed to the Spaniards by the two English traitors that Leicester had left in command.

Still, had it been otherwise, it is certain Drake would not have been allowed to go. Mary Stuart was under sentence of death, and the attitude of France was more menacing than ever. For two years Philip had been at work upon his Armada. His ports were crowded with its details; his storehouses were bursting with its furniture; and Walsingham at last was able to convince the Queen by a paper stolen from the very closet of the Pope, that it was upon her head the great engine was to crash. Her eyes were opened; and infected for a moment with the warlike spirit into which her people and her Parliament had lashed themselves, she ordered Drake to the coast of Spain.

It was no longer as a privateer that he was to act. He held the rank of Her Majesty's Admiral-at-the-Seas, and William Borough, the Comptroller of the Navy, was his vice-admiral. Four of the Queen's largest battle-ships and two of her pinnaces were under his command, and the London merchants committed to his flag ten fine cruisers, with the famous *Merchant Royal* at their head. Besides these he had six hundred tons of his own shipping, as well as some of the Lord Admiral's. In all, exclusive of tenders, there were twenty-three sail—five battle-ships, two first-class cruisers, seven of the second class, and nine gunboats large and small. With this fine force he was instructed to proceed to Cape St. Vincent, and by every means in his power to prevent the concentration of the several divisions of the Armada, by cutting off their victuallers and even destroying them in the ports where they lay. If the enemy sailed for England or Ireland, he was to hang on their skirts, cut off stragglers, and prevent a landing; and, finally, he was given a free hand to act against the East and West India convoys.

Elizabeth was in a resolute mood. Drake's ideas of naval warfare were developing a step further, and the Queen for the moment listened. He was beginning dimly to grasp that the command of the sea was the first object for a naval power to aim at. It was because he had not command of the seas that he had been unable to retain his hold of Cartagena, for the troops which should have formed its garrison were wanted to defend his fleet. Wiser for the lesson, his aim was now to crush the Spanish navy, and then in undisputed control of the sea to gather in his harvest. The opposition were thoroughly alarmed, and while Drake in hot haste was driving on his preparations, they left no stone unturned to get his orders modified. They tampered with his men, they whispered slanders in his mistress's ear, they frightened her with threats from abroad, they tempted her with offers of peace from Parma on the old disgraceful terms. For Walsingham, who, through thick and thin, was always at Drake's back, it was an unequal fight; with the staunchest of his party in disgrace for Mary's premature execution, he was single-handed against a host, and at last the friends of Spain prevailed. Early in April a messenger sped down to Plymouth with orders that operations were to be confined to the high seas. As Philip's ships were all snug in port, and could well remain there as long as Drake's stores allowed him to keep the sea, it was a complete triumph for Spain. But when the messenger dashed into Plymouth with the fatal packet he found the roadstead empty. Drake was gone.

In vain at the last moment a number of his sailors had been induced to desert; he had filled their places with soldiers. In vain a swift pinnace was despatched in pursuit; Drake had taken care no orders should catch him, and with his squadron increased by two warships from Lyme, was already off Finisterre, battling with a gale which drove the pinnace home. For seven days it raged and forced the fleet far out to sea. Still Drake held on in its teeth, and so well had he his ships in hand, that on the 16th, within twenty-four hours after the gale had blown itself out, the whole fleet in perfect order was sailing gaily eastwards past Cape St. Vincent.

Eastwards—for he had intelligence that Cadiz harbour was full of transports and store-ships, and on the afternoon of the 19th as he entered the bay he saw a forest of masts in the road behind the city. A council of war was summoned at once, and without asking their opinion he quietly told them he was going to attack. It was his usual manner of holding a council, but it took Borough's breath away. It shocked the old Queen's officer, and outraged his sense of what was due to his own reputation and experience, and the time-honoured customs of war. He wanted to talk about it, and think about it, and find out first whether it was too dangerous. And there was certainly some excuse for his caution. Cadiz stands on a precipitous rock at the end of a low and narrow neck of land, some five miles in length, running parallel to the coast. Within this natural breakwater are enclosed an outer and an inner port, and so cumbered with shoals and rocks was the entrance from the sea that no ship could get in without passing under the guns of the town batteries, while access from the outer to the inner port was only to be gained by the Puntal passage half a mile wide. Opposite Cadiz, on the other side of the outer harbour, was Port St. Mary, and within the Puntal channel, at the extreme end of the inlet, stood Port Royal. Both places, however, were so protected by shoals as to be unapproachable except to the port pilots. It was an ideal scene of action for galleys to develop their full capabilities. Two had already appeared to reconnoitre, and how many more there were no one could tell. Galleys, it must be remembered, were then considered the most formidable warships afloat and quite invincible in confined waters or calms. By all the rules of war, on which Borough was the first authority in the service, to attack was suicide; but Drake had spent his life in breaking rules. He did not care. The enemy was there, his authority was in his pocket, the wind was fair, his officers believed in him, and as the sun sank low behind them the fleet went in.

A scene of terror and confusion followed. Every ship in the harbour cut its cables and sought safety in flight, some to sea, some across the bay to St. Mary's, some through the Puntal passage to the inner harbour and Port Royal. To cover the stampede ten galleys came confidently out from under the Cadiz batteries. All was useless. While the chartered cruisers swooped on the fugitives, the Queen's ships stood in to head off the advancing galleys as coolly as though they had fought them a hundred times before. In a few minutes the English Admiral had taught the world a new lesson in tactics. Galleys could only fire straight ahead; and as they came on line abreast, Drake, passing with the Queen's four battle-ships athwart their course, poured in his heavy broadsides. Never before had such gunnery been seen. Ere the galleys were within effective range for their own ordnance they were raked and riddled and confounded, and to the consternation of the Spaniards they broke for the cover of the batteries. Two had to be hauled up to prevent their sinking, the rest were a shambles, and nothing was now thought of but how to protect the city from the assault which seemed inevitable. Hardly any troops were there: a panic seized the population; and Drake was left alone to do the work for which he had come.

Beyond the batteries the fleet anchored with its prizes, plundering and scuttling with all its might, till the flood came in again. Then all that remained were fired, and by the flare of the blazing hulks as they drifted clear with the tide, Drake moved the fleet into the mouth of the Puntal channel out of range of the batteries. He himself took up a position seawards of the new anchorage, to engage the guns which the Spaniards were bringing down from the town and to keep off the galleys; for as yet the work was but half done. In the inner harbour lay the splendid galleon of the Marquis de Santa Cruz and a crowd of great ships too big to seek the refuge of the shoals about Port Royal, and at daylight the *Merchant Royal* went boldly in with all the tenders in company. Then, in spite of the labours of the past night, the plundering, scuttling, and burning began again. Outside, the galleys were making half-hearted demonstrations against the English anchorage, but they were easily kept at bay. By noon it was all over, and Drake attempted to make sail. In the past thirty-six hours he had entirely revictualled his fleet with wine, oil, biscuit, and dried fruits. He had destroyed some twelve thousand tons of shipping, including some of the finest vessels afloat, and four ships laden with provisions were in possession of his prize crews.[1] It was enough and more than enough. But the wind would not serve, and all day long he lay where he was, in sight of the troops that were now pouring along the isthmus into Cadiz. Again and again the galleys attempted to approach, and every time Drake's broadsides swept them back before they reached their effective range. Vainly too the Spaniards strove to post guns near enough to annoy the fleet. Nor did the struggle cease till at midnight a land-wind sprang up, and brushing from his path the galleys that sought to block the way, Drake made sail. By two o'clock he had cleared the batteries and was safe outside without losing a single man. Boldly enough then the galleys gave chase, but, unfortunately, the wind suddenly shifted completely round. Drake at once went about, and

[1] In the official report the Spaniards admit the loss of twenty-four ships valued at 172,000 ducats. This, it would seem, was all they dared tell the King. Duro, *Armada*, i. 334, where the report is printed.

the galleys fled in most undignified haste, leaving the English fleet to complete its triumph by anchoring unmolested in full view of the town.[1]

Such an exploit was without precedent. The chivalry of Spain was as enthusiastic in its admiration of Drake's feat of arms as it was disgusted at the cumbrous organisation which condemned it to inactivity. A whole day Drake waited where he was to try and exchange his prisoners for English galley-slaves, but getting nothing but high compliments and dilatory answers for his pains, on the morrow he sailed. There was no time to lose. By his captures he had discovered the whole of Philip's plan. Out of the Mediterranean the divisions of Italy, Sicily, and Andalusia were to come and join the headquarters at Lisbon, where the Grand Admiral of Spain, the Marquis de Santa Cruz, was busy with the bulk of the Armada. At Cape St. Vincent was the road where ships coming out of the Straits waited for a wind to carry them North, and there he had resolved to take his stand, and fight everything that attempted to join Santa Cruz's flag in the Tagus.

Such light airs prevailed that it was not till the end of the month that the fleet reached the road. By that time its water was exhausted, and as every headland was crowned with works commanding the anchorage and the watering-places, Drake at once saw he must take them. In his usual off-hand way he summoned his council, and told them over the dinner-table what he was going

[1] For the whole action cf. the authorities collected in *Camden Soc. Mesc.* vol. v., with the plan put in by Borough at his court-martial, *S. P. Dom.* ccii. 14. i.

to do. It was more than the vice-admiral's dignity and caution could endure. In high dudgeon he returned to his ship, and in the midst of a gale which suddenly arose and drove the fleet to the north of the Cape, he indicted a long and solemn protest, not only against the contemplated operation, but against the unprecedented despotism with which Drake was conducting the whole expedition. Borough, though no doubt jealous of Drake, certainly believed he was doing nothing beyond his right and duty. He felt he had been attached to the expedition as the most complete sailor in the kingdom, and he valued and deserved his reputation. In the scientific knowledge of his art he was unrivalled, and he was the only officer in the service who had fought and won a purely naval action. No one, therefore, can fairly blame him for resenting the revolutionary manner in which his commander was ignoring him in contempt of the time-honoured privileges of the council of war. Drake in his hot self-confidence thought otherwise. As he rode out the gale under the lee of St. Vincent, and the tempest howled through his rigging, once more there fell upon him the shadow of the tragedy which could never cease to darken his judgment. Already, in Cadiz harbour, he had thought his vice-admiral too careful of his ship when the shot were flying; and now he saw in him another Doughty sent by the friends of Spain to hang on his arm. "In persisting," he told Lord Burleigh, "he committed a double offence, not only against me, but it toucheth further." To his embittered sense the querulous protest was a treasonable attack on his own authority, and in his fury he brutally dismissed the old admiral from his command, and placed him under arrest on his flagship. In vain the astonished veteran protested his innocence, apologised, and made submission. Drake would not listen. The ring of the headsman's sword upon the desolate shores of Patagonia had deafened his ears to such entreaties for ever.

Two days later he was back in Lagos Bay, landing a thousand men for an attempt upon the town, but in the evening, after vainly endeavouring to induce the bodies of cavalry which hovered on their line of march to come within reach, the troops re-embarked, reporting the place too strong to be taken by assault. Such reports were not to Drake's liking. It was no mere cross-raiding on which he was bent, but a sagacious stroke that was essential to the development of his new ideas. To get the command of the seas it was necessary that he should be able to keep the seas, and for this a safe anchorage and watering-places were necessary. In default of Lagos, strategy and convenience both indicated St. Vincent road for his purpose. It was commanded by forts, but that did not deter him; and resolved to have his way he next day landed in person near Cape Sagres. On the summit of the headland was a castle accessible on two sides only. The English military officers declared that a hundred determined men could hold it against the whole of Drake's force. But he would not listen; it commanded the watering-place, and he meant to have it. Detaching part of his force against a neighbouring fort, which was at once evacuated, he himself advanced against the castle, and at the summit of the cliff found himself confronted with walls thirty feet high, bristling with brass guns and crowded with soldiers. The garrison had just been reinforced by that of the evacuated fort, and to every one but the Admiral the affair was hopeless. He attacked with his musketeers, and when they had exhausted their ammunition, in the name of his Queen and mistress he summoned the place to surrender. In the name of his lord and master the Spanish captain laughed at him. Whereupon Drake, more obstinate than ever, sent down to the fleet for faggots, and began piling them against the outer gate to fire it. So desperate was the resistance that again and again the attempt failed. For two hours the struggle lasted. As fast as the defenders threw down the fire, the English piled it up again; and in the midst of the smoke and the bullets the Admiral toiled like a common seaman, with his arms full of faggots and his face black with soot. How long his obstinacy would have continued it is impossible to say, but at the end of the two hours the Spanish commandant sank under his wounds, and the garrison surrendered. Daunted by a feat which every one regarded as little short of a miracle, the castle and monastery of St. Vincent, together with another fort near it, capitulated at the magician's first summons, and left him in complete possession of the anchorage to water the fleet undisturbed.

Having fired the captured strongholds, and tumbled their guns over the cliffs into the sea, Drake returned to the fleet to find the sailors had not been idle. Between St. Vincent and a village some nine miles to the eastward which they had been ordered to burn, they had taken forty-seven barks and caravels laden with stores for the Armada, and destroyed between fifty and sixty fishing-boats with miles of nets. The tunny fishery, on which the whole of the adjacent country chiefly depended for its subsistence, was annihilated. For the time Drake's work on the Algarve coast was done, and having watered the fleet and fished up the captured guns, he sailed for Lisbon.

His own idea had been to land there and smite Philip's preparation at its heart, but this the Government had expressly forbidden.[1] Still he hoped that the havoc he had made and the insults he had put on the Spanish coasts might goad Santa Cruz to come out and fight him. For three days he lay off Cascaes in sight of Lisbon, threatening an attack and sending polished taunts to the Spanish admiral. He offered to convoy him to England if his course lay that way; he took prizes under his very nose; with his fleet in loose order he sailed up to the very entrance of the harbour; but though seven galleys lay on their oars watching him from the mouth of the Tagus, Santa Cruz would not move, and Drake learned at last how deep was the wound he had inflicted. Philip's organisation was completely dislocated. The fleet at Lisbon was unmanned. Its crews had been shattered in Cadiz harbour, and the troops that were intended for it had been thrown into the defenceless city under the Duke of Medina-Sidonia with orders that while Drake was on the coast not a man was to be moved. All thought of an attack on England was given up. It was even doubted whether by straining every nerve it would be possible to save the homeward-bound fleets from the Indies. The Italian squadrons were ordered to

[1] Drake to Walsingham, June 2nd, 1587. *S. P. Dom. Eliz.* ccii. 7.

land their troops at Cartagena, and Philip hoped that by forced marches across the peninsula they might possibly arrive in time for Santa Cruz to sail before it was too late. Every one else looked on the convoys as doomed. For Drake, having assured himself that Santa Cruz could not stir, and that England was safe for a year at least, resolved to make for the Azores and wait for the prey that had so narrowly escaped him the year before. On the third day of his stay off the Tagus he took advantage of a northerly gale to run for the anchorage at St. Vincent, which he had made his own and where he intended to water and refresh for the voyage. There, huddled under the lee of the Cape, was found a fresh crowd of store-ships which he seized. For nine days he lay there rummaging the ships, taking in water, and sending the men ashore in batches to shake off the sickness with which, as usual, the fleet was attacked. Every day new prizes fell into his hands, and ere he sailed he had taken and destroyed forty more vessels and a hundred small craft. On May 22nd he put to sea, and as the news spread a panic seized every commercial centre in the Spanish dominions. Half the merchants in Philip's empire saw ruin before them: the whole year's produce both of the East and West Indian trade was at Drake's mercy; and no one knew how Spain with its resources already strained to the utmost would survive the shock.

Whatever might have been the result had these fears been realised, destiny seemed to have decided that in the Channel should be played the last great scene. Drake had not been two days out when a storm struck his fleet and scattered it over the face of the sea. For three days it raged with extraordinary fury. Drake's own flagship was in dire peril, and when the heavens cleared only three of the battle-ships and half a dozen smaller craft were together. Not a single merchant-ship was to be seen, and the *Lion*, Borough's flagship, on which he was still a prisoner, was missing too. Before leaving St. Vincent, Drake had told Walsingham that he ought to have at least six more cruisers to do his work properly, and now two-thirds of what he had before were gone. Still he held on, hoping to find some of the missing ships at the rendezvous in the Azores. On the morning of June 8th St. Michael's was sighted, but not a sail had rejoined the flag except the *Spy*, one of the Queen's gunboats, with the captain and master of the *Lion* on board, and they reported that the crew of Borough's ship had mutinied and carried him home. Then in the depth of his disappointment Drake's fury blazed out anew. His fierce self-reliance and fanatic patriotism had taught him to see a traitor in every man that opposed him, and the bitter experience of his life-long struggle against the enemies of his country and his creed could bring him but to one conclusion—Borough was the traitor who had ruined the greatest chance of his career! A jury was empanelled, the deserter tried for his life, found guilty, and condemned to death.

It was little good except to relieve the Admiral's anger. The splendid opportunity was gone: the fruit of his brilliant exploit was snatched from his lips; for even had the remnant of his fleet been less shattered than it was, the great convoys were beyond its strength. The only hope was to hurry back to England and beg for reinforcements to fight Santa Cruz for the life-blood of Spain.

Yet ere he sailed there was a consolation at hand. As he lay waiting for his shattered squadron to close up, fuming at traitors, and marvelling at the inscrutable will of Heaven, the dawn of June 9th lit up the gray sea and showed him a huge carack in the offing. On a smart breeze he gave chase. The carack kept her course, but as Drake drew near began displaying her colours nervously. Drake made not a sign in reply, but held on till he was within range. Then on a sudden, with a blaze of her ensigns and her broadside, the *Elizabeth Bonaventura* told the stranger what she was. Two of Drake's squadron threw themselves resolutely athwart-hawse of the enemy, and the rest, plying her hard with shot, prepared to run aboard her towering hull. But ere they closed, her flag fluttered sadly down, and the famous *San Filippe*, the King of Spain's own East-Indiaman, the largest merchantman afloat, was a prize in Drake's hands. Well might he wonder now at God's providence, as with lightened heart he sailed homeward with his prize. For not only was it the richest ever seen in England before or since, not only was its cargo valued at over a million of our money, but in it were papers which disclosed to our merchants all the mysteries and richness of the East India trade. It was a revelation to English commerce. It intoxicated the soberest capitalists; and they knew no rest till they had formed the great East India Company, to widen the gap which Drake had opened and to lay the foundation of our Indian Empire.

CHAPTER X

IN QUEST OF THE SPANISH ARMADA

STILL burning for action, on June 26th Drake reached Plymouth with his splendid prize, and after a fortnight was spent in getting her up to Saltash and overhauling the priceless cargo, with a chest full of jewels and the bill of lading he hurried to Court hoping to dazzle the Queen into giving him the orders he so ardently desired. But all was in vain. Not all Drake's temptation, nor all that Burleigh and Walsingham could urge, availed to stir the obstinate mood into which Mary Stuart's execution had plunged her. She would attend to nothing but the funeral. Burleigh and all the war-party were still in disgrace for having stolen the serpent from her bosom, and she stubbornly shut her ear to all who did not speak of peace. Drake's exploits promised to wreck the whole negotiations, and he was pitilessly reprimanded. So far from being allowed to assist him, Burleigh was set to write despatches assuring Parma that the Admiral had exceeded his instructions and was in disgrace. Orders were sent down to pay off his ships, and the hunger which his small beginning, as he called it, had only whetted, had to go unappeased.

Still he might have been content, for the actual havoc he had made was but a little thing beside its moral effect. Not only had he taught English seamen to despise the dreaded galleys, but in the rank and file of Philip's host he had planted a terror against which it was vain to struggle. That a Lutheran heretic could so prevail against the army of God could admit, as men thought then, of but one explanation, and that the Church made haste to foster. Drake was a magician. He had sold his soul to Satan for a familiar by whose aid he worked. In his cabin was a glass in which was shown him the fleets of his enemies and all that passed on board: he could count their crews and watch their movements; and like the Norse witches of old, by some dark bargain he had bought the power to garner the winds and loose or bind them at his will.[1]

Let no one underrate what all this meant. He cannot read aright the history of that time, who fails to grasp how such a personality could oppress the imagination. Sorcery was then as real as sin, and men moved and breathed and thought in an atmosphere charged with magic. Nor was this all. If the superstitious fishermen that manned Philip's ships shuddered before a new devil, the romantic chivalry of Castile had found another Roland. For the crews, to fight was madness; for the captains, surrender was no shame. To the King his name was a torment. The grandees looked cold disdain when it was uttered. The Pope mocked at him, and said Elizabeth's distaff was keener than Philip's sword. He invited a lady to go upon the water, and she protested she dare not for fear the Dragon should come and take her from her sovereign's arms. Philip

[1] Navarrete, *Documentos Ineditos*, tom. 81, p. 245. *Hist. MSS. Com. Rep.* XII. iv. 252.

banished her from the Court, and smarting under the scourge redoubled his activity. But still he had to feel what foreign critics were saying openly, that in England was a man before whom his Armada might be not invincible and his crusade a disgrace.

Yet he relaxed no fibre; nor did Drake. Forbidden to strike Philip abroad, he turned his animosity against the traitor he saw at home. Though the Government refused to carry out the sentence of death, Borough was brought before a court-martial. Charge upon charge Drake heaped implacably on his head, and confounded him with crowds of witnesses too eager to win the great Admiral's favour. Yet to his indignation and astonishment the Court refused to convict the prisoner of treason. More they could not do. It was impossible for them not to find that the veteran who seventeen years ago had so brilliantly defeated the Baltic pirates had lost his nerve; and so with clouded reputation in administrative employ, and once in command of a despatch-vessel, he fades from history moaning hopelessly over the charges which had broken his heart.

But to crush the man who, as he believed, had ruined his enterprise was not enough for Drake's energy. For twenty years he had never ceased a day to do and dare against Spain, and he was not likely to be still at such an hour as this. If the Queen would not make war he was determined to do it on his own account. As he looked round him for the best method of pursuing his life-long quest, his eyes could not but turn on the abortive project of five years ago. It had been in his mind some time. As he lay off Lisbon in May he had ascertained that the Portuguese were expecting him to land with Don Antonio in his company, and for the rest of the time he was on the coast he had been carefully preparing his ground by conciliating them in every way he could. Don Antonio, over head and ears in debt, was still hanging about the capital ready for anything that would release him from the clutches of his creditors. Black John Norreys, Drake's old brother-in-arms, was there too, out of employ and in disgrace for presuming to try and save the English arms in the Low Countries from Leicester's incompetence. The capture of the great carack had set the merchants' mouths watering for the Indian fleet, and everything seemed ripe for a repetition of the great king-making project. The only difficulty was the Queen. But Drake had every ground for a comfortable faith in her love of tortuous political moves. He knew too at what value to set her reprimands, and, moreover, he had at Court a new friend more powerful and eager than Hatton himself. The young Earl of Essex, the son of his old patron, was just now in the first flush of his favour and his passion for adventure. He had just been caught in an attempt to escape to the seat of war in the Low Countries, and brought back to play games with his fond and lonely mistress. Smelling traitors now with every breath, Drake pursued his intrigue in such deep mystery that only here and there his workings showed on the surface. Still there can be no doubt that he suggested to the forlorn young truant a new way of escape. The coaxing of her favourite and the temptations of her little pirate were always hard for the Queen to resist; and as she found her apologies to Spain accepted and the negotiations for peace going smoothly once more, opposition in high quarters seemed to disappear. Don Antonio received a thousand pounds to pay his more pressing debts; ships began to collect at Plymouth; the carack was ordered to be sold, that the merchants interested might refit their vessels; and by the end of October Drake had formed a syndicate to provide the fifty thousand pounds which was required of him as a privateer by way of caution-money. Everything promised well for his new war under Don Antonio's flag, when all at once Elizabeth was confronted with the fruit of her folly in not having permitted Drake to return at once and complete his work. Walsingham's brilliant financial operations by which he had got the King of Spain's bills protested at Genoa were in vain, for the safe arrival of the great convoys had restored Spanish credit, and stung at last from his patience Philip found himself rich enough to indulge in an outburst of energy that surprised both friends and foes. Regardless of the season the Armada was to sail ere the year was out, and England, after the years of warning, was taken by surprise. Once more the country was tossing in a fever of warlike preparation. The navy was to be put on its war-footing, and Drake was summoned to headquarters to take counsel for the safety of the realm.

In endeavouring to appreciate the strategy of this time, for which Drake and Sir John Norreys must be held responsible as respectively the naval and military chiefs of the staff, it must be remembered that England was threatened by three separate invasions at the same moment. In Spain was the Armada; in Flanders was Parma with an army of thirty thousand of the finest soldiers in Europe, with adequate transport and a small fleet to convoy them; while the Border was in peril from the Scots. Any two of these dangers, or even all three, might combine; but the best intelligence led to the belief that Parma meant to join the Scots, while the Armada seized Ireland or the Isle of Wight as a base of operations against the west or south. In view of this information and the fact that the Queen, still obstinately clinging to her hope of peace, would only openly sanction a defensive war, there is little fault to be found with the English naval dispositions. A fleet under Lord Henry Seymour, with Sir William Wynter and Sir Henry Palmer as flag-officers, was to watch Parma in the narrow seas and to act in concert with the Dutch, who were blockading the Spanish Netherland ports. To command in chief Lord Howard of Effingham was commissioned Lord High Admiral, with authority to invade the Spanish dominions. For this purpose he was to be in command of the main fleet, with John Hawkins and Martin Frobisher for his flag-officers. For Drake was reserved the high rank of Lieutenant to the Lord High Admiral, an office which seemed designed to give him as full a control over the war at sea as the lingering feudality of the constitution would allow to a commoner, however great his professional capacity. By virtue of the office he became President of the Naval Council of War, and as the Lord Admiral's deputy, could exercise all the powers of that officer's commission at the head of an independent command. With this in view his little fleet of privateers was reinforced from the Thames and Portsmouth dockyards with four battle-ships, a cruiser, and a couple of smart gunboats. His division was thus raised to thirty sail, and the plan of action seems to have been that while Howard guarded the

Channel, Drake was to inflict a counter-blow somewhere. It was given out that he was once more bound for the Spanish Main, and volunteers flocked to his flag. His real destination was kept a profound secret, but we cannot doubt what it was. For at Christmas time a spy was reporting to Burleigh that Don Antonio might easily be restored to his throne; and in January the instructions of the commissioners, who were going over to Flanders to treat for an armistice, were modified so as to forbid Portugal or Don Antonio being included in the negotiations. In the light of this extraordinary piece of statecraft, Elizabeth appears hardly so single-hearted in her struggle to keep the peace as some have thought her; but she was at least consistent. Her darling policy had been all along to do her brother-in-law grievous bodily harm without committing a breach of the peace, and Drake had ever been the weapon that most nicely fitted her hand. She could not believe that Philip's patience was at last exhausted; and under Don Antonio's flag she thought, like Celia in the play, to make herself invisible and "catch the strong fellow by the leg."

Such was Drake's mission, as on January 3rd he went down to Plymouth to hoist his flag. It was fitly borne by the immortal *Revenge*, than which no ship was ever more gilded with the romance of war. His old friend, Thomas Fenner, was his vice-admiral in the *Nonpareil*. His rear-admiral was Captain Cross in the *Hope*. Edward Fenner commanded the *Swiftsure*, his fourth battle-ship; and Will Fenner's flag flew over his cruiser the *Aid*. Beside the royal ships rode five splendid merchantmen of London, perfectly found, as the Londoners always were. The rest were west-country craft belonging to himself and to his own and his wife's relations and friends. All outward-bound vessels had been stayed, and Drake could pick his crews from the flower of the English marine, who flocked to his flag in numbers, it was said, sufficient to man two hundred sail. True, half-crews only had been sanctioned, but to this foolish piece of economy Drake paid no attention. Regardless of all but his end, he manned his fleet with its full complement, and when the time came sent in the bill without a word.

While Drake was thus busy with his expeditionary force Howard covered Plymouth, and watched for the Armada off the Land's End. But he watched in vain. The seas were still free: the winter campaign seemed a false alarm; and Howard, about the middle of January, was recalled to the Thames, where, in spite of his protests, half his crews were paid off. Early in February came a new alarm, and Howard was once more ordered to man his ships and put to sea. But while the Lord Admiral and Seymour were thus distracted with orders that changed with every new report from Spain and every new turn of the negotiations, Drake, except when he practised too hard with his big guns, was not meddled with. Men said he would still sail; and Philip, trembling for his reconstructed fleet, left no stone unturned to get him stopped. Crofts, his pensioner in the English Council, even went so far as to tempt the Queen's cupidity with a scheme for his disgrace and the confiscation of his wealth.[1] Essex, too, was made to suspect that the Admiral meant to play him false and treat him as he had treated Sydney; and as his preparations approached completion, Drake grew more and more anxious. Nor was

[1] Crofts to the Queen, Feb. 21st, 1588. *S. P. Spain*, xxii. 33.

it without cause. For ere the month was out Crofts had prevailed, the commissioners for peace went over to Ostend, and the Plymouth fleet was stopped.

From Howard downwards the sailors were in despair. All through March he and Seymour were compelled to waste their resources with a naval demonstration off Ostend, in support of negotiations which they knew instinctively were but a trick. Drake was kept idle at Plymouth, and it was thought that when the moment came he could not possibly be ready. We can see him fuming up and down the Plymouth Hoe as he looked down on his half-dismantled ships, growing fouler and fouler as they chafed at their moorings. We hear him swearing and praying by turns as he gazes seawards for a trace of the gunboats he has sent to Finisterre for intelligence that will convince his mistress that the peace-negotiation is only a trick to disarm her. Still the weeks went by, and nothing was done till, at the end of the month, he received orders to get ready for sea. Then from the depth of his disgust he poured out one last appeal to the Council. During his impatient striding up and down the Hoe, the true theory of naval warfare, of which he had already a dim perception, had been growing clearer in his teeming mind, and in his acknowledgment of the fresh orders he, for the first time, distinctly formulates the idea of getting command of the sea. He fully grasped that the invasion was to come from Parma in the Netherlands; but no less perfectly he perceived that its feasibility hung upon the possession of the four seas.

"If Her Majesty," he wrote, "and your Lordships think that the King of Spain meaneth any invasion in England, then doubtless his force is and will be great in Spain, and therein he will make his groundwork or foundation whereby the Prince of Parma may have the better entrance, which in mine own judgment is most to be feared. But if there may be such a stay or stop made, by any means of this fleet, in Spain, that they may not come through the seas as conquerors (which I assure myself they think to do), then shall the Prince of Parma have such a check thereby as were meet." With deep apologies he urged the folly of keeping so large a fleet blockading Parma, and craved that his own division might be strengthened, that he might go and seek the enemies of God and Her Majesty wherever they were to be found. Still struggling to give clear utterance to the idea with which his genius was in travail, he went on only to confuse it with the moral effect of offensive operations till his passion altogether overcame his argument, and he told how three hundred English flags with the red cross had been made in Lisbon. "Which is," he bursts out, "a great presumption, proceeding from the haughtiness and pride of the Spaniard, and not to be tolerated by any true natural English heart."

His appeal had some effect, for on its heels came a despatch from the commissioners at Ostend, telling how Parma had admitted that warlike preparations were still going on in Spain, though he vowed they were only against Drake. So an order came down that he was to write to the Queen direct and tell her how strong her fleet ought to be to carry out his ideas, and how he proposed to distress the fleet which was assembling at Lisbon. To the second question he would give no direct reply. Traitors were too thick about the Queen; and he told her it depended upon the intelligence he got on the way, and the temper of his force when he got it to sea. "The last insample at Cadiz," he growled in the bitterness of his anger, "is not of divers yet forgotten, for one such flying now as Borough did then will put the whole in peril." As to the strength of her fleet, "God increase your most excellent Majesty's forces daily," said he; but with four more navy ships and sixteen merchantmen that were fitting out at London, he declared himself ready through the goodness of his merciful God to answer for the Armada—or even, as the advantage of time and place in all martial actions was half the victory, he offered to sail as he was, and let the reinforcements follow. Surly as was the tone of his answer, the Queen was delighted and sang his praises everywhere. But still she could not make up her mind to loose her growling dog. A fortnight later one of the gunboats came in with intelligence that showed the Armada was on the eve of sailing, and Drake, in an agony of impatience, hurried the captain up to Court, urging again as a matter of life and death that he should be allowed to go. The answer was a summons to town. Elizabeth was now thoroughly alarmed, and no sooner was the sailor's rough eloquence heard ringing in the Council-chamber than the Queen's purpose was at last made firm; the eyes of the Government were opened to the great idea, and Howard with every ship that had three months' victuals was ordered to join Drake in the west.

Golden weeks had been wasted. How grave the peril was, those only grasped who had to face it; and Drake knew that if the Armada was once allowed to sail, the England he loved so well was at the mercy of God. It was not till May 23rd that Howard reached

Plymouth. Drake was already there. In the morning light at the head of sixty sail he put out to greet the Lord Admiral, and then, as they met, went about with his whole division and escorted his delighted chief into port. There the combined fleet watered, and on May 30th, as an easterly breeze sprang up, the two Admirals put to sea in loyal concert, to try if there yet were time to strike the weapon from the hand that threatened their country's life.

It was no easy task that was before them. Over the sea came uncertain sounds of preparations so vast that no one could doubt any longer where Philip's right arm was. But where and how it would strike was still uncertain. It might be destined for Ireland or for Scotland. It might be meant to seize some English port. It might be under orders to join Parma, or to act with the Guises from France. It might come north about by the Orkneys or directly up the Channel, and to watch one route was to leave the other open. Even if the Armada's course were divined aright, the wind which brought it must throw the English to leeward; and to reap the advantage of our superior gunnery, on which the only hope of victory lay, the weather-gage was essential. For the defending force it was a situation as difficult as that which outwitted Nelson himself, and one well designed to force home Drake's idea of the command of the seas. From the first Drake had seen the strategic and tactical disadvantages of attempting to cover any of the threatened points. To prevent those threats ever being developed was his plan; and to effect this he saw he must go boldly out and lie to windward of the enemy's port of departure. Once there he felt that even if they dared come out he could so handle them as they put to sea, and so harass their advance, that nothing but a broken remnant would ever reach the British coasts. As far as men could see it was the only chance; and hoping against hope that there might yet be time, the Admirals lay the course for Finisterre. But their cup was not yet full. Ere they were clear of the Channel the wind veered to south, and began to freshen to a gale in their very teeth. With it came over the deserted seas a solitary merchantman, which announced that ten days ago she had seen the whole Armada stretching westerly from horizon to horizon on a northerly wind. It was too late. The wind which had stopped the English fleet would bring the enemy, and there was nothing to do but to stand off and on where they were. For six days the gale continued, veering slowly, but in spite of it the fleet held its ground. On the seventh day it was blowing harder than ever straight from the west, and then fearing to be driven so far to the leeward as to uncover Plymouth, the discomfited Admirals put back.

The disastrous situation which for months had been haunting Drake's dreams was at last waking truth. Still the Spaniards came not; and a ray of hope brightened the gloom as Drake, with the instinct of a born strategist, divined what had happened behind the waste of storm. He felt that what had been seen was only a move to a rendezvous at Corunna. There was still time to strike. The conditions were indeed more favourable than ever. After the gale the Armada must take time for a final concentration, and backed by Hawkins, Frobisher, and Fenner, he persuaded Howard to try again. Their determination was at once announced to the Council. But day after day the westerly gale continued to rage. Such a summer had never been seen. Every attempt to get to sea failed, and ere the fleet got free, to Drake's utter dismay there came a peremptory order from the Queen absolutely forbidding the manœuvre. In Drake's absence some one had frightened her back into the old and futile methods; and with a sharp reprimand for rashness, Howard was directed to cruise between Spain and England, and to water nowhere but on his own coasts.

Had the Queen in her perverseness wished to destroy her fleet as well as paralyse it she could hardly have given more fitting orders. With a sullen growl the Admirals obeyed. It was June 19th before they could get out, and in three days they had to put back for want of victuals. It was not till the next evening that the provision-ships arrived, and it was fortunate they did. For the same night came news that eighteen Spanish ships had been sighted off Scilly. Without a moment's hesitation a few stores were flung on board the fleet, and leaving the victuallers to follow, it pushed out to sea again on a fresh north-easterly breeze to cut off the straggling squadron. But again, ere they were out of the Channel, the wind chopped round to south-west and stopped further progress. It was the wind to bring the Armada; at any moment its sails might appear; so while Howard stood off and on in mid-channel, Drake, with ten ships and four or five gunboats, made a sweep down to the Bay to feel for the enemy there, and to retard their advance if he found them stealing up the French coast. Howard had a fixed idea that the first intention of the Armada was to join hands with the Guises in some French port; but though Drake bowed to the Lord Admiral's superior political information, from the first his instinct told him the movement was only exhausting the fleet to no purpose. By this time it was known that two more stray Spanish squadrons had been hovering about Scilly. But for days no sign of them had appeared. One of them had even been sighted bearing for Spain, and Drake divined the rest. As though he had indeed been shown the truth in a magic mirror, he knew that the late gales had broken up the Armada and that it must be painfully reassembling in the ports about Finisterre. As he paced impatiently the deck of the *Revenge*, gazing out over the still desolate sea, he saw in Vigo, and Bayona, and Corunna a confusion of shattered rigging, and heard the muttering of landsmen sickened of the sea and raw crews demoralised with failure. Into the midst he pictured himself bursting like a thunder-clap, and in a storm of fire and iron completing the ruin which Heaven had begun. In a week he could endure it no longer. Victuals were running out: the crews, on half rations, were falling sick; and as every day some poor fellow was flung overboard they began to lose heart. June came to an end; and then Drake rejoined the Lord Admiral to try and prove to him with all the force of his eloquence how the Lord had once more in His mercy delivered the enemy into their hand.

All was in vain. Howard, loyal to his fatal instructions, would do nothing but stretch out his fleet like a net across the mouth of the Channel, and patrol his front and flanks with gunboats. In desperation Drake reduced his reasons to writing and sent them home, imploring to be allowed to go at least a little nearer to Spain, in order that, even if he were wrong, and the Spaniards were already on their way, he might still have some chance of getting to windward of them before they entered the Channel. Shaken at length by his lieutenant's vehemence Howard ventured to stretch his scruples so far as to advance the line outside the Channel; and there on the afternoon of the 7th a fresh northerly breeze came up behind them. To Drake it was the very breath of the Lord, and before his passionate conviction Howard at last gave way. Half the fleet had but a few days' provisions, but, as Drake pointed out, if they returned for more the other half would be just as bad, so without more ado away they went for Finisterre. Long and low the tempter must have laughed to himself as they flew before the wind. If it only held, orders or no orders, for bare life's sake his scrupulous commander would be compelled to revictual from Philip's storeships. No man ever watched the wind more anxiously as next day sails began to shake ominously. Yards were braced round, bowlines were strained more and more, yet league by league they neared the goal. On the 9th Ushant was eighty leagues behind, but their labour was in vain—a south-wester was blowing in their teeth. To proceed was impossible, to stay was starvation, and in open wonder that God should have sent a south-wester, Drake confessed that retreat was the only course. So for the third time the great Armada escaped. Helpless and wind-bound it had been lying in Corunna Bay, at the mercy of the fireships and great guns of a fleet to windward. But now the wished-for wind had come to release it, and on the 12th, as Howard's exhausted fleet reappeared off Plymouth, the Duke of Medina Sidonia, in the fulness of his strength, put out to sea untouched.[1]

Not an English sail was there to see or hinder, and ignorant of the jeopardy in which they stood, Howard and Drake set every hand to work that their fleets might be ready to renew the attempt together the moment the wind was fair. A messenger sped to Court for permission, and this time, so convincing had Drake's memorandum proved, it was not refused. By the 19th they were almost ready—sick had been landed, crews were reinforced, the scanty stores allowed were on board—when suddenly they were astounded with the news that the Armada was off the Lizard. The tables were completely turned. By the south-west wind on which Medina-Sidonia was advancing, the English were shut in port and caught in the very same trap which Drake had meant to be the destruction of the enemy. Clever as he was he had not guessed the whole truth. He could not tell that the wandering squadrons were merely some stray ships that had kept on bravely in spite of the gales to the rendezvous at Scilly; he could not tell that the bulk of the Armada, more faint-hearted, had never passed Finisterre, but had taken shelter

[1] For this forgotten movement see *S. P. Dom. Eliz.* ccxii. f. 9. *Reasons offered by Sir Fr. Drake*, etc., f. 10. *Considerations proposed by Mr. Thos. Fenner*, etc. (wrongly dated July 4th), f. 57. Thos. Cely to Lord Burleigh, f. 80. Howard to Walsingham (postscript), f. 82. *Relacion de . . . Pablo de Arambur que . . . fué con dos zabras en seguimiento del Armada inglesa*, Duro, ii. 213.

weeks ago ere irreparable damage was done. After all his scheming and strife with friend and foe he was taken by surprise at last, and the Armada had reached the Channel without one English gun to say it nay.

CHAPTER XI

THE BATTLE OF GRAVELINES

THE old story goes that the fatal tidings found Drake on Plymouth Hoe playing bowls with the great officers of the fleet. He was out-manœuvred, the surprise was complete, but he did not stir or start. Of all that gallant company none knew so well as he all that the tidings meant. But in the jovial face, ruddy and clear-eyed as ever, there was no sign of the anxiety beneath. His fleet lay huddled in port, at the mercy of the Spanish fireships: there was not a moment to lose; but hurry would spoil all, and he would not budge. There was time, he said, to play the game and beat the Spaniards too. Born leader of men that he was, his genius for a timely bravado taught him the value of such a speech to quiet panic for the tremendous effort he saw at hand. For there was but one thing that could save the fleet. It must be got out of harbour before the morning, and then the almost hopeless struggle would begin. Well-nigh incredible as the feat may seem, it was accomplished. During the night, in the teeth of the wind, nearly all the fleet was warped out. With the utmost difficulty and consummate seamanship, on Saturday morning Rame Head was weathered, and in rain and mist the fleet began beating to the westward close along the coast to try and steal the wind from the Spaniards.

So thick was the weather that it was not till the afternoon that the fleets had sight of one another. Then to the masthead of the great *San Martin* the Captain-General of the Ocean Sea raised the blessed standard of the Crusade. On one side was the crucified Redeemer, on the other the Holy Mother. Three solemn guns boomed out, and as the sound rolled through the great Armada every man fell down and prayed to Christ for victory against the enemies of His Holy Faith. Then, too, along the misty shore where Drake's men toiled were muttered unrecorded orisons without ceremony and without prostration. They were crusaders too. Their faith was as deep, their worship as devout; but the Puritans hauled on their bowlines as they prayed. There was need for them to aid the Lord. The Armada was edging inshore very close; but closer still the arch-enemy of the Holy Faith crept on in the wet mists to complete his manœuvre. Fainter and fainter sank the wind and heavier fell the rain, as Drake wrestled with the weather. The chance of success was desperate indeed. In those days the best of ships could not sail within six points of the wind, and at every tack the countless sails on the horizon loomed clearer and closer as they crept on before the dying breeze. Yet on the completion of the movement before the dawn, he saw the fate of England hang. On the morrow Plymouth would be at the Spaniards' mercy, unless the English by getting the wind could fall upon their rear and force them to leeward of the threatened port. Night fell with the work still far from done, and hour after hour in the darkness were heard the cries of the boatswains and the wail of the rigging as the ships went about continually. But when the curtain of the night was lifted it was upon a glorious scene for England. A little west of Looe was the great Armada lazily advancing still, and seawards right in the wind of it were forming the two divisions of the English fleet. Out of Plymouth was sailing another squadron, which, as it passed boldly across the Spaniards' front, fired on the leading ships, and then going about joined the bulk of the fleet to windward. Then in loose order Howard and Drake bore down. Anticipating that Sidonia with his superior numbers and tonnage would take the offensive, they had resolved to attempt nothing but to harass his advance and cut off stragglers. But no sooner were they within range, than they recognised how hard was the task before them. Far from attempting Plymouth the Armada received them in the crescent formation, and continued its way up channel unmoved by their fire. So well were the Spaniards disposed, and such splendid order did they keep, that it was impossible for the English to come to close quarters without danger of losing the wind. Compelled to fight at long range it was in vain that they directed the whole weight of their metal upon the port division of the Spaniards' line, and tried by crowding it upon the centre to break up their unassailable formation. Sidonia's best advisers knew well their weakness, and Drake, too, saw it at a glance. "The Fleet of Spaniards," he hurriedly scrawled on the despatch that was sent to warn Seymour, "are somewhat above a hundred sails, many great ships; but truly, I think, not half of them men of war." He was right. The Armada was not a fleet of warships free to take the offensive as it would. Philip, still less than Elizabeth, had grasped the importance of commanding the sea. Instead of being a fighting machine of overwhelming power, his great Armada was in reality the convoy of an unwieldy mass of transports and storeships; and Sidonia's orders were to escort it straight to Margate, and not to risk an action with the combined English fleet till he was safely at his destination.

But though nothing could tempt Sidonia from his defensive attitude, most of his flag-officers were all on fire to fight. Their chivalry was slow to learn the lesson which Drake so suddenly had taught both king and subject. They could not believe that their long unquestioned supremacy at sea was gone; and at last Don Juan Martinez de Recalde in the *Santa Anna*, who commanded the port division, as he felt himself being forced into the centre, was galled into turning on his foes. In a moment he was surrounded by the van of Drake's division. The *Revenge* herself was there, conspicuous with an extravagant pennant and a banner on her mizzen, and fighting almost at grappling distance. Only one vessel could get to Recalde's support, for as his flag-officers bore up to relieve him, they were received by an overwhelming fire from Howard. So hot grew the fight, and so smartly was ship after ship brought to bear on the isolated Spaniards, that at last Sidonia himself was compelled to come up into the wind to save them. Howard signalled to sheer off; and for the rest of the day, in order to cover Recalde while he refitted his shattered flagship, Sidonia kept the English busy with manœuvres for the wind.

So ended the memorable Sunday. Little enough had been done except to show the superior activity of the English ships, and the greater rapidity of their fire. They found that under courses only they could outsail the Spanish galleons, and could easily pour in three broadsides to the enemy's one. Still the Armada was almost untouched. At nightfall Sidonia, having re-formed his battle-ships, was pursuing his way up channel after the storeships and transports as majestically as ever, and the seamen were grumbling that the onset had been more coldly done than became the credit of the English navy. Drake saw that the tactics of the past day would never do. It was clear that Sidonia meant to avoid an action, and act purely on the defensive till he had joined hands with Parma. Had the contest begun off the Spanish coast, as Drake had desired, a series of harassing engagements might have succeeded; but now there was no time. The two Spanish forces were but a few days asunder, and at all hazards they must not be permitted to unite. Ere the light faded, therefore, the flag of council was flying on the *Revenge*, and when the flag-officers were assembled, Drake, and those who supported him, urged that a great effort should be made without a moment's delay. But, right or wrong, Howard lacked the daring of the seamen. He could not bring himself to risk the fleet, —the only hope of his unready country; and indeed the risk was great. Every ship grappled by a Spaniard was doomed; and lacking the old hands' implicit faith in the power of English seamanship to elude the danger, Howard would not give way. With a discretion for which he cannot be blamed, he resolved to continue the tactics of the past day. Until his whole force was concentrated by a junction with Seymour and Wynter in the Straits of Dover, and by the arrival of the reinforcements he expected, he was determined not to fight a general action, but to rest content with retarding the Spanish advance till the gales returned for their discomfiture.

Still bent, however, as we may well believe, on having his own way, Drake could console himself that as vice-admiral of the combined fleet the van of the pursuit was his. The captains had orders to follow his cresset light, and with the wind freshening and veering to west-north-west he led the chase of the Spanish stern-lanterns. Under easy sail he clung upon their heels, till in the middle of the night he was aware of some strange sails near him, which were bearing a different course, and impressed as he was with the way the Spanish fleet had been handled during the day, he was suddenly seized with the idea that they were weathering him in the dark. Something at any rate was wrong, and with characteristic decision he felt he must get to the bottom of it. In accordance, therefore, with the naval practice of the time in such cases, and nothing loath perhaps to mislead the commander he could not direct, he immediately extinguished his light, and accompanied only by his tender, stood with the strangers. Having satisfied himself they were but a few German merchantmen seeking convoy, Drake at once gave up the chase and let them go. Meanwhile, however, his manœuvre had thrown the English fleet into confusion. Howard, taking the Spanish light for Drake's, held on his course. Four or five of his division did the same.

The rest, at a loss what to do, struck sail, and when day broke the *Revenge* found herself alone, save for a huge galleon drifting within a few cables' length of her. It was none other than *Nuestra Señora del Rosario*, the flagship of Don Pedro de Valdes, captain-general of the Andalusian squadron, and one of the finest officers in the fleet. In trying to assist Recalde he had fallen a victim to Sidonia's rigid formation, which, though well enough before the wind, too often led to collisions if any manœuvre was attempted. Drake at once summoned her to surrender, but though she had lost her bowsprit and foremast, she had hitherto beaten off every assailant, and honourable conditions were demanded. The English Admiral was in a hurry; he wanted to overtake his division; he was Drake, he said, and had no time to parley; and at the sound of the great name Valdes struck without another word. He and forty of his officers were taken on board the *Revenge*, where they ransacked mythology to find adequate compliments for their captor's prowess and generosity. Proud of his reputation, Drake loaded them with a princely hospitality, while his officers took possession of their treasure of some fifty thousand ducats. Then sending the galleon with a prize crew into Dartmouth, he went on his way with his prisoners to overtake the Lord Admiral.

By this time Howard was far ahead. So closely indeed had he followed the lights which he took for his lieutenant's, that at break of day he had found himself amongst the rearmost ships of the Spaniards. Knowing as we do Drake's persistent methods of getting his way when persuasion failed, it is impossible not to suspect him of a hope, if not of an intention, of entrapping Howard into a general action by his wayward manœuvre. But in that case he was disappointed, for so light was the wind, and so scattered the fleet, that it was four in the afternoon before the ships had closed up, and by the time Drake arrived on the scene the Lord Admiral had extricated himself without fighting, and an attack was no longer possible. All that Howard and Hawkins had been able to do was to take possession of the flagship of the Guipuzcoan squadron, which had been shattered by an explosion in its powder-magazine and abandoned. Otherwise the Armada was as strong as ever; and to make matters worse, Sidonia had taken advantage of the respite to reorganise his force. Taught by yesterday's experience the weakness of the crescent formation in face of so nimble an enemy, he united and reinforced his two wings, and formed them into a rear-guard, while he himself, with the galleasses and the transports and storeships, formed a van division. In this formation, as Monday drew to an end, the Armada was lying becalmed off Portland. Towards sunset the wind had died away entirely, and thus another precious day was gone and nothing done.

The moon rose clear and bright, and, mirrored on the glassy sea, the two fleets lay facing each other scarce a cannon-shot apart. The English were spread motionless in a long broken line at the mercy of the enemy's oared ships. It was the hour for the galleasses, and from the extreme right of the Armada where they lay, in all the pomp of their swinging oars they came striding over the moonlit waters upon an isolated vessel of the English. But at that moment troubled patches began to darken the shining surface of the sea: sails began to shake themselves for the coming work; and in a few minutes a smart breeze had robbed the galleasses of their prey. The doomed ship was saved; but the wind was coming from the north-east and the Spaniards had the weather-gage. The English at once led off with a dash straight inshore to try and get round the enemy's right. Sidonia, to parry the movement, stood in too with his motley division trailing after him, and signalled to the rearguard to follow. But no sooner was the Admiral's move taken up than the English fleet was round again and standing to the eastward on the opposite tack threatening to turn the Spaniards' left. By the change of direction Drake, with the starboard division, was now leading, closely followed by the Lord Admiral, and as they strove to pass to windward, the fighting rearguard of the Armada, which by Sidonia's inshore movement was now on its left, turned suddenly upon them. A hot action was the immediate result, and at a distance so close that the English ships were in constant danger of being boarded. One was only saved by the Lord Admiral boldly offering its assailant a chance of boarding his flagship, and he himself was in imminent danger till he was rescued by Drake's vice-admiral. To abide such an attack from the windward was rank heresy to the new tactics of Drake's school, and nimbly disengaging, the whole English fleet sprang its luff and ran large out to sea, to deprive the enemy of his chance of boarding. After it the Spaniards' rear-guard laboured, straggling more and more every length it advanced. Far from being defeated, as the Spaniards thought, the English were fast doing what had seemed almost hopeless. Sidonia's attempt to follow their rapid doubling had broken his solid phalanx to pieces. Far away to windward was his flagship toiling after his chasing rearguard: farther still behind him were the galleasses vainly struggling with the current in the Portland Race to keep their positions in his wake; and scattered confusedly over the sea were groups of transports and victuallers trying to regain the shelter of the battleships. As the day advanced every hour improved the English prospects, for the wind was going slowly round with the sun. For a time they contented themselves by luffing up continually to deliver broadsides on their pursuers and tempt them to straggle farther; but at last the wind had veered far enough to give them the weather-gage, and with one accord they swept relentlessly upon Recalde's flagship. Aware that it was still trying to recover from the punishment its boldness had met with on Sunday, the whole Spanish rearguard bore up one by one to the rescue, and that with the greatest devotion, for each ship as it came into action was the victim of a concentrated fire. Sidonia in like manner was bearing up with his squadron too when his flag-captain pointed out to him that away to leeward the transports and storeships were in sore trouble. The galleasses, it was evident, could not protect them. They had given up the attempt to follow their leader's flag, but their attention was fully engaged by the *Triumph*, the finest vessel in the English navy and the flagship of Frobisher. He had got left to leeward too, but was now bravely entertaining the galleasses, while Edward Fenton, in the *Mary Rose*, and several chartered ships of Drake's and the London squadrons were improving the occasion around him. Seeing the danger the Duke signalled to his immediate following to keep on to where the fight raged round the crippled *Santa Anna*, while he himself went about and bore down to support the galleasses and protect his more defenceless charges. The *Triumph* was now in serious danger; nor could the English seamen, having once exposed the Armada's vulnerable point, permit it to be covered again without a blow. Moreover, the continual closing up of the scattered Spanish rearguard had rendered the struggle round Recalde too equal to suit Drake's ideas. Once more he and his fellows disengaged, and Sidonia suddenly found himself alone between the transports and his rearguard, with all the English battle-ships bearing down upon him. Having reduced the *Santa Anna* to such a wreck that Recalde was compelled to transfer his flag, they were bent on treating as hardly the Captain-General of the Ocean Sea. With all the old-world chivalry of Spain the Duke shortened sail to await the attack, and for an hour endured alone the whole fire of his enemy, as ship by ship passed by and plunged its broadside into the great *San Martin*. From its fortress of the poop, from its castle of the prow, from its *plaza de armas* in the waist, the splinters shivered and split till the water poured in through the shot-holes, the rigging hung in ruin, and the holy standard of the Crusade was rent in twain. So they left him and passed on to support Frobisher, and gather the fruit of Drake's bewildering tactics. Under all sail the Spanish battle-ships toiled to the rescue, but when, as the sun sank low, Sidonia had once more gathered up his flock into a roundel, there were many that laboured sorely, and three were gone.

It was a poor enough result for a hard-fought action. The English powder was spent, another day was gone, and still the Armada looked almost as formidable as ever. Yet the effect was deeper than it seemed. With no little alarm the Spanish officers had been shown the mobility of a fleet formed line-ahead, and its power of concentration on weak points. It was the first dawn of those modern tactics which Blake and Monk were to develop and Nelson to perfect, and both sides recognised the great fact. The Spaniards' hearts sank as they saw how ill adapted were their floating fortresses for the new situation, and a bolder purpose inspired Lord Howard. On the morrow, though active operations were confined to twice compelling the Armada to stop its advance and form line of battle to its rear, a most important step was taken. After the morning cannonade had ceased, for some hours the English ships were seen tacking hither and thither in strange disorder; but at last, out of the confusion, four distinct lines developed themselves and bore down on the wondering Spaniards to compel them once more to shorten sail and haul to the wind. Yet not a shot was fired; and no sooner was the Spanish battle-array completely formed than the four lines twisted back like snakes and left it untouched. Vexed to be so fooled, and convinced that the English meant only to delay him till the fine weather should break, Sidonia resolved to turn no more till the spires of Calais rose in sight. But he was wrong. Reinforced and supplied anew with powder, Howard at last felt justified in pushing home an attack. Three days had passed since the fleets first engaged; three days more would bring the Armada to its goal, and although its advance had been well delayed there was still no sign of a return of the tempestuous weather. It would no longer do to watch the face of the skies. If the Armada's power was to be broken it must be by the hand of man.

By this time they were abreast of the Isle of Wight, and its proximity emphasized the necessity for prompt offensive action. From his exalted prisoner Drake had been able to learn that in certain contingencies the island was to be occupied in order to provide a harbour of refuge for the Armada, and it was clear the great effort could no longer be delayed. Nor could a better battle-field be wished for than the spot the two fleets had now reached. As they lay off Sandown Bay between the Island and Selsea Bill, the Armada had in its lee a whole network of shoals stretching from the Owers to Spithead, and a vigorous attack from windward promised either to press it in confusion amongst the intricate channels off Portsmouth, or if it attempted to weather the Bill and so regain the open sea, to drive it on the Ower bank. Such at least would seem to have been Drake's idea; and it was probably to increase the energy of the attack that Howard had consented to divide the fleet into four divisions. He himself commanded the first and Drake retained his own as the second, while the third and fourth were given respectively to Howard's two flag-officers, Hawkins and Frobisher. To attack Sidonia's serried ranks as they were was but to court defeat; and in council of war it had been resolved that during the night six merchantmen from each division, by engaging at four different points, were to loosen the Spaniards' formation and prepare it for the attack of the battle-ships.

So Drake meant to wrestle his great fall with the Spanish power. It was a well-laid scheme, and had the wind held, the Armada might never have sailed another league to the east; but as it chanced, the wind fell so light that the detailed merchantmen were unable to carry out their orders, and as the morning of the fourth day broke serene and calm, the only sign of movement was where some of Hawkins's vessels were seen trying to tow themselves alongside the *Santa Anna* and another crippled Spaniard. Well-nigh helpless with their wounds, they had drifted from the ranks and seemed an easy prey. Three galleasses came sweeping to the rescue in all their majesty of oar and sail; but it was not to victory. With dogged courage the leading ships of Howard's and Frobisher's divisions were towed by their boats to meet the floating castles, and were soon tearing and shattering them with chain-shot and a hail of balls. The boldest of the Spanish rearguard struggled in to support, and at last Sidonia signalled for a general action. It was the feast of San Domingo, his patron saint; a light breeze had sprung up in his favour, and with banners flying and trumpets braying, convinced at last that its hour of victory had come, the Armada formed in line of battle. The holy standard rose aloft and the tide of battle turned. The disabled galleasses were got out of action, and Frobisher and Howard, cut off and surrounded, seemed doomed to meet no better fate. The rest of the fleet were standing away as though to save themselves and desert their comrades, and the Spaniards felt certain of their prey. Still, like the heroes of some Homeric fight, the isolated admirals fought on, dealing destruction around and clearing themselves with their boats from every ship that attempted to lay them aboard. In wonder and admiration the Spaniards still pressed closer till the wind began to freshen. Then at last they learnt the meaning of the strange tactics of the enemy's two starboard divisions. While Howard and Frobisher were holding the Spaniards over the brink of the pit, Drake and Hawkins had laboriously secured a vantage-ground from which to thrust them down; and ere Sidonia had well grasped the sudden jeopardy in which his whole fleet was placed, the two kinsmen, with half a gale of wind in their swelling sails, crashed in upon his left. The charge was irresistible. The amazed weather squadrons gave way, crowding in upon the centre, and forcing the whole Armada to leeward. In vain did Frobisher's persecutors turn. Howard was free now, and added the weight of his consorts to the confusion with a new attack. The mouth of Portsmouth roads yawned on Sidonia's lee, as though to engulf his Armada; down came the holy standard of his crusading king; in its place flew the signal to close up in a roundel; and so, to the indignation of his fighting admirals, he hastened to make his escape, and inclining away to the coast of France, saved his fleet from the Ower banks.

"A hot fray," wrote Hawkins, "wherein some store of powder was spent, and, after all, little done." The change of wind had saved Sidonia. Drake and his kinsman knew their movement had failed, but Howard celebrated it as a victory. As the two fleets lay becalmed next day, some two miles apart, on the poop of the *Ark Royal* he knighted Frobisher and three noble kinsmen of his own who had fought their ships at his side.

They were the heroes the poets sang, and well they deserved the praise. But though Hawkins was made Sir John with the rest, for the cool head that had planned the fight and so nearly destroyed the enemy with strange tactics, in which bards could see no meaning, there was no reward. Frobisher, the intrepid explorer, who knew nothing of naval warfare, even accused the rival who had eclipsed his fame of cowardice. When he heard Drake had taken Don Pedro de Valdes and his great galleon, he swore the man that had brought England to all her trouble had deliberately lagged behind to reap the reward of his comrades' courage. The story is sad to tell, but let it be the glory of Elizabeth's captains that in the heat of their jealousy and quarrels they never forgot the sacred cause she personified.

The last chance of destroying the Armada before it reached its destination was now gone. All Friday the two fleets lay within sight of each other, till in the evening the breeze got up from the south-west, and the Armada began the last stage of its adventurous voyage. So splendid was the order it kept before the wind, that though the English still dogged its heels, no attack was attempted; and ashore spurs pressed deep and beacons flamed, faint hearts sank and brave men trooped together, as almost untouched the great host drew to its goal. On Saturday afternoon it was passing Calais—six leagues more and it would reach Dunkirk—when suddenly it came to anchor. Completely surprised, the English so nearly overran the enemy that they only saved the weather-gage by boldly anchoring within gun-shot to windward.

It was a solemn hour, as Seymour and Wynter

from the Channel fleet joined with the rest on Howard's flagship in momentous council. As the great German historian has said, the fortune of mankind hung on the balance. When we consider the difficulties of the navigation even for a single ship, the projecting headlands of the Channel, the intricate currents, the precarious winds, it is impossible not to admire the brilliant manner in which the great Armada had been brought to its destination. In spite of the greatest seamen of the age, in spite of the incubus of a cumbrous convoy, it had reached within arm's-length of its goal, and amid the hurried tramp of the gathering levies arose loud railing at the English captains for their failure. The Spaniards themselves could hardly credit their success. They were dispirited with their losses: to leeward lay the unknown terrors of the North Sea; to windward was a horror worse than all. For there they knew was El Draque, busy brewing in his ships, as they rode so peacefully at their anchors, the devilry for which he had let them come so far. Such thoughts at least the terror of that name conjured as the night fell. Nor were they very wrong. Who first thought of it, none can tell. Indeed it matters little. Since that terrible night in Vera Cruz harbour twenty years ago, and later, when the devil-ship blew Parma's legions to atoms on the Antwerp boom, the device was in every man's mind. It was a remedy hardly fit for Christian men to use. Yet, at all hazards, the Armada must be dislodged. The Dutch fleet, which had been blockading Dunkirk, had been compelled by its necessities to retire and leave Parma free to come out. At any moment the weather might serve, and in a few hours the great army might have passed the Channel. It was on Sunday the council of war met, and ere it separated the grim resolve was taken. That night the tide would serve, and Sir Henry Palmer sped to Dover for the means which, for this supreme moment, had been collected there; but no sooner was he gone than there were some that grew impatient and saw that he could not return in time to catch the tide. It would not do to risk the loss of another day; there were ships in plenty with the fleet. Drake offered one of his own for the sacrifice,[1] and seven more were quickly chosen. As the night closed in dark and moonless, a presentiment of impending doom disquieted the great host to leeward. The awful tragedy of Antwerp was in every mind, and hither and thither flitted launches patrolling to windward of the Spanish anchorage in nervous expectation. Midnight had passed, the night was at its blackest, and the rushing tide swirled dark and angry through the crowded galleons as they lay labouring, each with two anchors out. In the depth of the gloom whence the flood was sweeping with the wind, the English lights were twinkling peacefully, till a sudden flare obscured their brightness. Then another and another burst out, and glowed and grew till eight flaming masses reddened the night, and sped forward with wind and tide upon the terror-stricken Armada. Such a sight man's eyes had never seen. What wonder if a panic seized the Spanish fleet? There was no time to weigh. In reckless haste cables were slipped and cut, and like a herd of stampeding cattle, in mad confusion the tide swept the

[1] See a MS. account of Drake's claims against the Government, preserved at Nutwell Court.

great fleet away, crashing ship on ship through a tangle of writhing cables.

What had happened could not be seen from the English decks. As the fireships passed over the enemy's anchorage other fires seemed to rise; but it was not till the gray dawn broke that they knew not a single Spanish ship had caught fire. In its chief object the terrible stroke had failed, but in its lesser aim it had succeeded entirely. The Armada was dislodged, and the bulk of it was seen still flying in disorder to the north-north-east. The *San Martin* had hove-to, and with a little group of galleons about her was firing guns and signalling to recall her panic-stricken flock. Nearer still the most splendid of the four galleasses was seen, shattered with a collision, struggling with oars and foresail to get under the guns of Calais Castle. Then Howard weighed. Every arrangement had been made for completing the work the fireships had left undone, and the whole fleet was ready to fling itself into the confusion in one last cast to fight till it could fight no more. The Lord Admiral was to lead, Drake was to follow, and Seymour to come last. But there was more need for rapid action than any one had dreamed, for no one expected that the whole eight fireships could have missed. All sail was crowded on. Seeing her signal disregarded, the *San Martin* was making for the flying Armada, and trying to rally it off Gravelines lest it should be swept to leeward of Dunkirk, and already the finest ships were going about to form upon her in the old unassailable order. The rest were luffing more and more to clear the Flemish sands, and perhaps, under cover of the gathering rearguard, to tack and return with the tide to their anchors. Everything hung on whether the attack could be pushed home before the enemy had formed. It was a question almost of minutes, and in that supreme moment the Lord Admiral was found wanting. For him the magnificence of the labouring galleasse was too tempting a lure. Unable for all his courage to comprehend the new tactics, his only idea was to pluck his enemy feather by feather, and as he came athwart Calais he turned aside. For the sake of capturing a vessel already out of action he risked the last chance of destroying the whole Armada. One tremor of irresolution and all might have been lost. But there was none. It was the hour for which Francis Drake had been born, and without a glance at his commander, in all his heroic directness of purpose he led the fleet onward.

Nearly the whole of the captains obeyed his lead, as in grim silence the *Revenge* bore down straight for the group in which the *San Martin* towered. Not a shot was fired till almost within pistol-range her bow-guns were let fly at Sidonia. Luffing immediately she next plunged in her broadside at point-blank range, and with that deadly salute passed on enveloped in smoke. Close at Drake's heels were Frobisher in the *Triumph* and Hawkins in the *Victory*, and others, enough to deal with the *San Martin* and her friends; beyond was another group of giants trying to form on Sidonia's flank; on these Drake fell, and was lost again in a fiery cloud. Away to the starboard arose the roar of Wynter's guns and Seymour's, as in like manner they dealt with other knots of the disordered rearguard. On every side the last great battle in the west was raging with fury unexampled. After three hours Howard with the lingerers came up, and as he plunged into the heart of it with timely aid the fury of the fight grew fiercer still. Eyes had never seen the like. Old hands who had fought at Lepanto swore it was child's play to this. There was no thought of prize or quarter. As each galleon's fire ceased and she was seen to settle, another was taken in hand. Nor were the Spaniards less determined. Though the English gunners plied them three and four to one, though beside their nimble enemy they moved like logs, they would not hear of surrender. The commander of one galleon stabbed a man who was in the act of hauling down his colours. Drake's rear-admiral, Cross, sank another at his side with its flag still flying and its captain's defiance ringing in his ears. They were as short of ammunition as the English, but when their cannon-shot were spent they defended their splintered and leaking vessels with musketry, till frantic with the carnage mariners leapt overboard by scores to be picked up by the busy pinnaces of the English. Let the roll of those who for the livelong day endured the storm to save the rest from the sands be remembered by the conquerors. Leyva, Recalde, Oquendo, Pimental, Enriquez, and Francisco de Toledo, all were there and many more, the flower of Spanish chivalry, fighting to the last almost without power of resistance as the water rushed in beneath their feet. Drake's decision at the critical moment had had its effect. They had never been able to more than half form, but back to back, as it were, in little groups, they had to endure the resolute charges of the English who surrounded them, and for six hours the battle had continued.

Shuffled together, and almost unmanageable, the Spanish ships had by this time lost all semblance of the crescent formation. Of the forty vessels that had been cut off from the rest of the Armada, only sixteen were left at the last. Some had fled, one had sunk, another was sinking, and a third was derelict. About three o'clock it began to rain heavily, and the fire of necessity slackened. But for three hours more a desultory fight continued, as Sidonia took advantage of the lull to gather the shattered battle-ships together for retreat. No one cared to press them hard. The bulk of the Armada was already to leeward of Dunkirk. The English powder was almost exhausted, and the breath of the Lord of Hosts was now sweeping the Spaniards to utter destruction. With the rain the wind had changed and freshened. It was blowing dead on the treacherous coast of the Netherlands: the sea was getting up and breaking over the decks of the labouring galleons; and Drake, himself riddled with shot, was content to hang upon the skirts of the flying enemy till he should see them, one by one, engulfed in the sands on their lee.

All through the dark and blustering night he clung upon them and watched. With his triumphant fleet on their weather quarter, he knew that half of them dare not tack, and the rest could not for their wounds. Yet inch by inch the squally north-west wind drifted them nearer their doom. For such an hour Drake's whole life had been lived—the life he had lived for vengeance on the idolaters and England's enemy. His life had been one long tragedy, and now, as he gazed over the lee-bow of the *Revenge*, in grim exhilaration he waited for the terrible catastrophe. The night gave way to day, and still the Armada was striving fruitlessly with a wind as relentless as the demon at its heels. All day the agony was prolonged, and as the last hour approached with the waning light, once more the Spanish crews fell down in prayer—but now it was not for victory. Six fathoms had been sounded, then five, and God alone, the pilots said, could save the fleet. And then it was at the eleventh hour that Drake had to own that vengeance was the Lord's. Suddenly the wind dropped, and then sprang up from the south. With a cry of thanksgiving, the Spanish helms were thrust down: in a few minutes the Armada was standing northward out to sea; and once more by the act of God, at the very moment when his vengeance seemed accomplished, the enemy had been wrested from Drake's grip.

"We have the army of Spain before us," he wrote to Walsingham on the following night, "and mind, with the grace of God, to wrestle a fall with them. There was never anything pleased me better than the seeing the enemy flying with a southerly wind to the northward. God grant you have a good eye to the Duke of Parma, for, with the grace of God, if we live I doubt not but ere it be long so to handle the matter with the Duke of Sidonia, as he shall wish himself at Saint Marie Port among his orange trees. God give us grace to depend upon Him, so we shall not doubt victory, for our cause is good." So on the heels of the flying Armada he cheerily announced its escape, and the grim resignation of the last words alone betray how sorely the Puritan's faith was tried. It was the last day of July. Yesterday Wynter and Seymour had been sent back to their old station in the Channel to watch

Dunkirk, and Drake, under peremptory orders from Court, had just sent his prisoners ashore. He was short of ammunition, he was weary watching, the long struggle seemed about to begin all over again, and yet no note was coming from him but hope and cheerful endeavour. For two days he and Howard with the rest continued the chase, but they were not destined to wrestle a fall. To the northward of the Dogger Bank the long-expected tempest came rolling up out of the west, and Drake knew, for a while at least, that the country was safe. Constantly threatened with attack, the Spaniards were holding resolutely to their old formation; and thus, as he had first caught sight of it, in the gathering darkness, with the wild storm-rack flying over it, Drake bid the Armada farewell, and through the rising sea ran for the coast of Scotland, to save himself from the wrath to come..

CHAPTER XII

DRAKE'S ARMADA

DRAKE was at the zenith of his fame. Though the battle of Gravelines, by some strange freak of destiny, is hardly known to the bulk of Englishmen, it was at least as momentous as Waterloo or Trafalgar, and the honour belongs to Drake no less rightly than the laurels of Copenhagen are Nelson's. Parma knew it well; and when Drake, leaving the Armada to the mercies of the westerly gales, struggled back through the tempest to face him ere the fine weather returned, he broke up his camp at Dunkirk and abandoned the enterprise. Howard knew it too; and when on his return from the chase he was suddenly summoned to Court, he was careful to furnish himself with a testimonial from Drake that he had behaved well to his lieutenant and taken his advice throughout. It was one of Drake's captains who was sent up with the trophies, and it was Drake's name with which Europe rang as the news of the victory spread. The remnants of the Spanish crews who escaped the rocks of the Orkneys and the iron cliffs of Connaught, came home to swear anew he was a devil and no man; and Medina-Sidonia slunk away to his home, to be tormented by urchins who cried under his windows, "Drake is coming, Drake is coming." It was only amongst the other captains that there was any disposition to deny him the glory. Frobisher's jealousy was made a focus of opposition by the friends of Spain; and Lord Henry Seymour, furious at having been ordered back to his blockade, begged to be relieved of his command if the Lord Admiral's lieutenant were going to serve in the narrow seas. Faithful and generous as Drake was to his followers, it is certain that in his masterful temperament there was something unbearable to those who were not content to walk in his train. While to his friends he was self-reliant, impetuous, and enthusiastic, to those whom his strong personality repelled he was egotistical, headstrong, and a braggart. Although it was never admitted, every one knew that his fortune had been founded in plunder; and in spite of his lavishness and his stern destruction of prizes whenever the interests of his country demanded the sacrifice, he was credited with avarice, and accused of dragging England into war to fill his pockets. Yet in truth he was greedy for nothing but vengeance and renown,—the renown of being hailed as the saviour of his country, the vengeance that was his religion.

He was now to give his ambition rein. As August passed and the fate of the Armada was known, scheme after scheme for the prosecution of the campaign was abandoned. The fleet was lying foul and idle in the narrow seas; the officers were quarrelling and splitting into factions; the sick and wounded were dying unpaid and uncared for in the streets of the seaport towns. The admirals did their best. Drake and Hawkins together founded the "Chatham Chest" for disabled seamen, but the Government, under the Queen's irresolution and parsimony, sank into apathy once more, and the country looked to Drake to say the next word.

It was with no uncertain voice that he spoke. It was imperative that something should be done before Spain ceased to reel under the blow she had received, yet nothing would the Government do. Drake went to Norreys. By the middle of September their plans were ripe, and those two knights whose brotherhood-in-arms had begun so darkly at the massacre of Rathlin, sent up to the Council a proposal that is fairly astounding. However much we may rub our eyes to see such a thing at a period that we are accustomed to look upon as one of the most glorious in our history, the fact is not to be denied. These two adventurous spirits, in the Queen's default, offered to form a great war syndicate to prosecute the struggle on which the national existence seemed to hang, and what is even stranger, their offer was accepted. All they asked of the Queen, or at least all they were granted, was a subscription of £20,000 to the syndicate, and the loan of six battle-ships. It was of course understood that the Government was in every way to facilitate their operations, and they were given power to press men and make requisition of provisions. A siege-train was also to be furnished out of the royal arsenals, and they were to be permitted to take into their pay thirteen veteran foot companies and six hundred horse from the English legion in the Low Countries. Norreys further received authority to pass into Holland, in order to negotiate for the co-operation of the Dutch. A joint expedition of the two great Protestant belligerents had been for years the dream of the English soldiers, and Norreys was so far successful that he procured the promise of ten companies and six warships. In England the idea was taken up with enthusiasm, and in Norreys's absence Drake was able to secure the support of most of the great seaport towns. Court and commerce came forward freely with money, and everything promised a triumph, ant success.

The general idea was a revival of the project for the liberation of Portugal in Don Antonio's name, but it was no longer on the modest scale of former years. Through all his life of stress and storm Drake had been the prophet of English nationality. His mission was to preach and demonstrate its innate strength, and now his victory had swelled his idea to its full development. It was no filibustering raid he had in his mind, but an imperial Armada as great as the one he had crushed, bent on conquest, and fit to show Europe that all that Spain could do was within the might of England.

Everything was to be ready by February 1st, and all the winter the reawakened war-spirit which Elizabeth had so long pent up surged round the two commanders. All that was warlike and adventurous in the country crowded tumultuously to their standard, and Drake renewed his relations of the Irish wars. The brothers oi Norreys were there, those Chickens of Mars who were regarded as the patterns of soldiership. There too came the truculent man of Monmouth, Sir Roger Williams, in his gilt morion and great plume of feathers, with all the crabbed pedantry and cool valour which seem to have given to Shakespeare the character of Captain Fluellen; and a host more besides whose reputations are long since dead, but whom under other names we may see to-day as clearly as when they lived quarrelling, fighting, and dying through pages of the Elizabethan dramatists. Nor was chivalry behind. At its head was the Earl of Northumberland, and Essex too had from Drake a renewal of his promise of a place, if at the last moment he could escape from Court.

The living force of England was loose at last, and fed by Drake's stupendous notions it began to develop an energy so formidable, that as it would seem the Government took alarm. The Queen began to assume more and more control over the preparations, and obstacles of all kinds arose. There were difficulties about the Low Country troops, and the Dutch too grew cold. Elizabeth would not sign the commissions; and when she did could not make up her mind to let them pass the seal. Courtiers tuning to her note began to back out and would not pay their calls, while the Earl of Northumberland was ordered to withdraw both his person and his subscription. Still the two knights persevered; but so straitened were they by the defaulters, and so protracted had been the preparations by the conduct of the Queen, that at last they had to apply to her for further assistance. She flew into a passion as a matter of course and tried to induce Lord Willoughby to take over the command. He had succeeded Leicester in the Low Countries, but though "he would not budge a single inch for all the devils in Hell" the Queen herself had managed to worry the heart out of him, and he begged to be excused from the service of so exasperating a mistress.

Perhaps there was something in the melancholy

dignity of the broken-hearted soldier's refusal that brought the Queen a touch of remorse; at any rate, in a fortnight the two knights were able to go down to Dover to hoist their flags. Still the forces they were to command were seriously below Drake's standard. They were already well into March, and the long delays had entailed a large expenditure in pay and freight to no purpose. Moreover, the siege-train had not been forthcoming, and there were also wanting the six Dutch men-of-war, seven companies of the English veterans, four of the Dutch, and all the cavalry, nor had they transports for more than twelve thousand men. But Drake was not to be beaten. He had set his heart on commanding a fleet as great as the Invincible Armada, and once free of the trammels of the Court he set about getting what he wanted in his own peculiar way. Till March 16th he waited at Dover, when there came sailing by a fleet of sixty-five Dutch vessels bound for Spain under passes from Parma. On these Drake pounced, captured them every one and carried them off to Plymouth, where the chartered cruisers were assembling. It was an extraordinary coincidence, as he told Walsingham with his tongue in his cheek, that the Dutchmen should have been passing the very day he sailed, especially as they happened to be exactly the class of vessels he wanted for transports; but be that as it may, the effect was electrical and volunteers flocked to Plymouth. His force was soon doubled, but as ill-luck would have it that only added to the generals' troubles. For a whole month they lay wind-bound, consuming their scanty store of victuals till they had barely a month's provisions left. It was not till April 6th that the wind was fair, and even then no sooner had they put to sea than they were immediately driven back. The case was getting desperate. To request a further supply was useless they knew, but they had still a strong card to play. Where entreaty failed a threat might yet succeed; so quietly pointing out to the Council that it was madness to sail with their existing resources, they bluntly announced that unless a reserve of stores was at once made ready to follow them, they would have to turn their army of twenty thousand men loose upon the country without a penny to pay them.

The letter found the Queen in a fine Tudor rage. Essex had suddenly disappeared from Court, Sir Roger Williams had not put back to Plymouth since the gale, and to her vexation in having to sanction a new requisition was added the conviction that the generals had connived at her truant pet's escape. Courtiers and messengers came spurring down the great western road, and pinnaces were fitted out to find the *Swiftsure* with which Sir Roger Williams had disappeared. With an edifying display of zeal for the fugitives' arrest the generals protested their innocence; but nevertheless on the 18th the fleet sailed without a trace having been discovered of Essex or Williams or the *Swiftsure.*

Drake's dream was realised at last. He had got Don Antonio snugly under his wing, and at the head of a fleet of one hundred and eighty sail he was bound for the coast of Spain. Not only is the armament noteworthy as being the most powerful that had ever left the English shores, but its organisation is of special interest as marking an attempt to introduce order into a naval force on military lines. Drake's admiration for the discipline and methods of soldiers no less than the example of the Armada was no doubt not without its weight. The fleet was divided into five squadrons, each of some seventeen ships and fifteen transports, and each under a colonel. These "colonels of squadrons" were the two generals-in-chief, Captain Thomas Fenner the vice-admiral, Sir Roger Williams, colonel-general of the foot, and Sir Edward Norreys, general of the artillery. At the head of each squadron was one of the Queen's battle-ships, and each squadron-colonel had on his staff a lieutenant-colonel, a "corporal of the squadron" or aide-de-camp, and a captain corresponding to the regimental captain-lieutenant. In the sixth Queen's ship sailed the rear-admiral Captain William Fenner unattached, as "marshal-general of the fleet." Associated with the two generals there was also a full military staff, including an intelligence department under a "master of the discoveries" with the rank of "lieutenant-colonel of the pinnaces." When all was over, the system was considered to have failed, but in truth it never had a fair trial, for so foul was the weather, and so hurried the whole affair from shortness of supplies, that the fleet was never once exercised at sea upon the new system.

Indeed so contrary fell the wind that a number of transports containing twenty-five companies of foot never weathered Ushant at all, and the generals had to continue the voyage in sadly crippled strength. Their instructions—for unhappily they were hampered with instructions—were first to destroy the remnants of the old Armada and the beginnings of the new one which were said to be lying in Santander and other ports on the north coast of Spain; then, if it could be done without too seriously compromising the Queen, they were to proceed to Lisbon and set Don Antonio on the throne which was supposed to be only waiting for him to take possession, and the whole was to be wound up, if Don Antonio consented, with an attack on the Azores and the capture of Philip's Indian and American fleets. Drake, however, had every reason for preferring his own intelligence to that of the Government, and his "discovery" department had reported that some two hundred sail had recently put into Corunna. Assured that it was the first-fruits of the new Armada, he made the foul weather an excuse for opening operations with a descent on that port. On April 24th the generals arrived in the road, and the same night by a skilfully combined naval and military operation seized the harbour and lower town almost without a blow. But little shipping was found. Four large galleons, however, including the *San Juan*, the flagship of Leyva, the vice-admiral of the Armada, were made prizes or burnt, and a vast quantity of stores found in the warehouses was taken and destroyed. But in spite of its success the expedition was in no pleasant plight. Though it was in full possession of the Lower Town and the surrounding country, the roadstead in which the fleet lay was still commanded by the guns of the Upper Town into which the Spanish garrison had retired, and to leave the anchorage in face of the prevailing foul weather was as desperate as to reduce the fortress without a siege-train. In backing out of her promise the Queen had no doubt intended that the expedition should be thus crippled, for in her womanly craving for peace she was still clinging to her fiction of a defensive war and did not wish her hopes to be compromised by offensive operations against Philip's territory. Still between attack and retreat the choice lay, and for men like Drake and Norreys such a choice is soon made. As a matter of course the more palatable cup was swallowed; and in order to divert the attention of the garrison from the fleet a siege was immediately ordered, and while his officers with great success raided the country round for cattle and such provisions as the warehouses did not provide, Norreys proceeded to do his best with mining and four insufficient guns. Till May 3rd under every difficulty the siege operations were continued, and then while Drake made a diversion with his boats in the harbour, with fierce impatience an assault was delivered. It was a mad attempt. So inadequate were the means at the generals' command that the breaches were found wholly impracticable, the storming parties were repulsed with heavy loss, nor had Drake with his flotilla any better success.

By this time, however, the weather was abating, and fearing to waste more time, the two generals determined to re-embark. But that was not so easy. Hardly was the resolution taken than the scouts reported that a force of some eight or nine thousand men having passed the bridge over the Mero River had occupied El Burgo and were intrenching themselves there in rear of the English lines. Thus finding themselves in the same position that Sir John Moore was to immortalise two centuries later, like him they resolved to give battle, and while Drake with five regiments held the trenches and siege-works, Norreys advanced with the remaining nine over the ground where Sir John Moore fell. Though in numbers inferior to the enemy, with such fury did he fling himself upon them that the intrenchments at El Burgo were swept with a rush; the bridge, which was only wide enough for three abreast, carried at push of pike; and in the evening Sir John Norreys, who, in doublet and hose, had been fighting all day beside his brothers pike in hand, marched back to Drake in triumph with the royal banner of Spain borne before him. The victory was complete and signal. The enemy had been pursued in open rout beyond the Mero, their loss had been very severe, the country for miles round was driven and ravaged, and on the 8th, having fired the Lower Town, the English re-embarked without hindrance by the light of the conflagration.

In spite of an alarming sickness which had broken out in the ranks, Drake as he led the fleet on its way was in excellent spirits. "We have done the King of Spain many pretty services at this place," he told Walsingham, "and yet I believe he will not thank us." Indeed the loss to the Spaniards in stores alone had been very great, while the idea of defeating in open fight a superior Spanish force intrenched on its own ground so intoxicated the commanders that, orders or no orders, they had no longer stomach for anything less than the invasion of Portugal. They had made up their minds that the Spaniards were so hard hit by the defeat of the Armada that it would be a mere promenade, and flinging to the winds the Queen's instructions about the north coast ports, they sailed for Peniche, a port under Cape Carvoeira some fifty miles north of Lisbon. The weather continued as bad as could be, and as the fleet toiled slowly southward sickness spread havoc through the crowded ships. On the fourth day out, however, they were gladdened by the sight of the long-lost *Swiftsure.* Essex and Williams had been down as far as Cadiz in search of Drake, and were now retracing their steps with a train of five prizes which they had picked up in Drake's old hunting-ground at Cape St. Vincent. The generals had stringent orders to send home Essex and put the Welsh captain under arrest as soon as they met, but stupid orders always sat lightly on both of them, and smoothing their conscience with the consideration that they could neither spare the *Swiftsure* nor Sir Roger Williams, who had been named as Norreys's successor in case of accident, they received the truants with open arms.

It was not until the 16th that they were able to anchor at Peniche. The sea was still running high, an ugly surf was breaking on the beach, but not a moment was lost. The boats were lowered away, and as they came under the fire of the castle the landing-parties with Essex at their head plunged into the sea and waded waist-deep to shore through the surf. The garrison sallied to resist them, but by a scientific flank movement from Sir Roger Williams they were compelled to fall back and suffer the rest to land without opposition. The town was carried by assault, and so rapid and complete was the success that the same night the castle opened its gates to Don Antonio. It was an encouraging omen, for the co-operation of the Pretender's partisans was, of course, an integral part of the general idea. Don Antonio had every assurance that the people would rise in his support, and that the appearance of the English before

Lisbon was to be the signal for its gates to be opened. The rest depended upon the rapidity with which the English commanders could anticipate the counter-movements of the Spaniards. Already much time had been lost, and the crippled force which had reached Corunna was decimated by casualties and disease. Waiting but a single day to land the handful of horse and refresh the men, they rapidly organised a flying column and Norreys on the 18th began to move on Lisbon. In the early morning light Drake took his stand upon some rising ground to bid God-speed to the little column, as in all their bravery of corselet and morion and the ensigns of their captains, pikemen and musketeers defiled before him. They were but fourteen weak battalions of recruits and one poor troop of horse; they had neither transport nor artillery; they hardly deserve the name of army, yet when we think of them with arms arust and breeches stained by the sea, tramping by to salute the Admiral, let it never be forgotten how grave a legacy they left to British arms, or of how long and glorious a procession they were the humble pioneers.

Having taken leave of the troops and left the sick and wounded in charge of a small force at Peniche, Drake himself went round to the mouth of the Tagus, where it had been arranged he was if possible to join hands again with Norreys, and support his attack on Lisbon with the fleet. Picking up a score or so of prizes by the way, on the 22nd he appeared in the Cascaes road. The inhabitants fled to the mountains at his approach, but on his sending ashore two of his Portuguese pilots to assure them that he was there in Don Antonio's name they returned, and although the Spanish garrison refused to surrender the castle, he took peaceable possession of the town.

Having thus strangely anticipated Wellington in securing a point for the re-embarkation of the army in case of accident, Drake sent out scouts to feel for Norreys. They returned with the startling intelligence that he was already quartered in the suburbs of the capital, and the Admiral at once set about fulfilling his promise. It was an operation of the gravest difficulty and danger. The Tagus from St. Julian to Lisbon bristled with forts, it was full of galleys, the navigation was hazardous, half the English crews were sick, and the masters declared solemnly in council of war that the attempt was madness. But Drake was deaf to their prudence. He had promised to meet his brother-in-arms at Lisbon, and formidable as he knew the batteries to be, with a good wind he believed he could run the gauntlet; while as for the galleys he could treat them as he had treated them before. In spite, therefore, of every protest two-thirds of the war-ships were told off for the service, and having equipped and armed them to the best of his ability he waited for a wind. The very next evening it began to change and Drake issued orders that the fleet was to weigh with the morning tide. But it was not to be. Ere the desperate signal was floating over the *Revenge* a messenger was standing before Drake announcing that Norreys was in full retreat on Cascaes.[1]

The expedition from which Elizabeth's soldiers had hoped so much had failed. The first of a long procession of exasperated officers, they had had to learn what good allies of the enemy were the apathy and suspicion of

[1] Drake's answer to the charges, *S. P. D. Eliz.* ccxxvii. f. 35.

the Portuguese. Though they had marched triumphantly through Torres Vedras to the gates of Lisbon, with everything falling back before them; though for three days they had occupied the suburbs of the capital in spite of every attempt of the garrison and the galleys to dislodge them; though at the entreaty of Don Antonio they had not stolen so much as a groat or a kiss, not a man had come to the Pretender's standard. Lisbon had refused to open its gates, Drake had not come with the guns and stores, the heat was making havoc in the disease-stricken ranks, and Norreys had resolved in deep disgust to pass on to the third part of the enterprise. As far as St. Julian's he was followed by a strong Spanish force proclaiming loudly, but at a respectful distance, that they had driven the English from the walls of Lisbon. Then all the knight-errantry of these old-world soldiers burst out in chivalrous excess. "Black John" sent a trumpet on the spot to the Spanish commander giving him the lie direct and challenging him to fight, army to army; the passionate Essex defied him to single combat or a party of ten to ten; and on the morrow at daybreak to make good their words they marched their broken forces back to the ground they had named,—to find nothing more valorous than a hastily-deserted camp.

Meanwhile Drake was garnering a harvest of prizes. In less than a week he had captured sixty sail of fine Scandinavian vessels which in spite of Elizabeth's warning were carrying contraband of war to Lisbon. On these all that disease and wounds had left of the army were embarked, and the Dutch shippers dismissed with an offer of corn for their pains. Now, too, arrived the reserve of victuals which the two generals had wrung from Elizabeth, and with it, to temper their satisfaction, a royal letter breathing such fury at their departure from her absurd instructions, and such peremptory resentment against Essex and Williams that it was thought wise to send the truants home before the voyage to the Azores was commenced.

On June 8th the fleet sailed to play its last card. The weather was still so unsettled that alternative instructions had to be issued fixing the rendezvous at Vigo or the Azores according as the wind fell south or north. On the second day out the wind dropped altogether and the fleet found itself becalmed off Cape de Espichel. Some twenty galleys which had come out of the Tagus to watch the retreat now plucked up heart to attack, and before the battle-ships could be towed to the rescue they had cut off four stragglers. Although they attempted nothing more it was but the first-fruit of disaster. For no sooner did the wind spring up than it rapidly developed into a southerly gale and scattered the fleet beyond hope. Having endured its fury for a whole week Drake, with some three squadrons which he had managed to collect, put into Bayona road. Finding no one there, he sailed at once for the Azores in search of his lost sheep; but hearing next day that Henry Norreys with his squadron had put into Vigo he returned. Still no tidings could be got of Edward Norreys and the rest, but, as it were to keep their hands in while they waited, Drake and Sir John Norreys took and burnt Vigo. But now so shattered were the ships and so reduced the crews—for the epidemic had never ceased its virulence—that it was clear a new resolution must be taken. As it was, the expedition was no longer fit for service, but unwilling, now that he was in complete command of the sea, to abandon the chance of a great blow at the Spanish trade, Drake proposed with a score of the best ships and the pick of the men to organise a cruising squadron and proceed to the Azores while Norreys took the rest home. The idea finding favour, instructions were given accordingly to weigh at once for the Bayona road in order to carry out the reorganisation, but hardly were the anchors up than a westerly gale of extraordinary fury caught the fleet and threatened it with destruction. Though by a magnificent display of seamanship Drake managed to get the bulk of it out to sea, the damage done was crushing, and in despair he held on with his storm-torn flock to Plymouth. On July 1st he staggered into the Sound, and on the morrow came Norreys in as evil a case as himself. As for the lost squadron it was not heard of for weeks afterwards. But Thomas Fenner was with it and others of Drake's men. So when it did come back no one was surprised to hear that it had reached the Madeira Islands and plundered Porto Santo.

CHAPTER XIII

THE LAST TREASURE-HUNT

THE Portugal Voyage marks the decline of Drake's star, and yet, when the effect of his Armada is compared with that of Philip's, it must seem a little strange that most of his contemporaries should have considered it so great a failure. The Spaniards, as the seamen boasted, "did not in all their sailing round about England so much as sink or take one ship, bark, pinnace, or cock-boat of ours, or even burn so much as one sheepcote on this land." Drake and Norreys had failed in their object too: the mortality amongst their men had been appalling; but when all was said and done they could point with pride to an exploit which, ere they had taught Europe the rising strength of England, men would have called incredible. In the ten weeks of their adventure they had not only destroyed the nucleus of a new Armada, but they had taken and burnt two Spanish ports; they had beaten one army of the King of Spain in the field and had made another run; they had marched a week through the heart of his territory; for three days they had insulted the gates of his second capital; they had captured nearly a hundred sail in his waters; and all this in spite of weather, sickness, and politics enough to ruin three such expeditions. Sir Roger

Williams told the Queen the expedition had done the King of Spain more harm and England more honour, service, and safety than all her expenditure in the Netherlands. But though at first she sent down to her two knights a glowing letter of thanks, and told them they had done all that valour and generalship could yield, yet she could not forgive them for not having wasted more time than they did in the north of Spain. Had they been less loyal than they were to her instructions, there is little doubt, so paralysed was the Spanish power, that the partisans of Don Antonio would have been able to deliver Lisbon into his hands. Nevertheless both generals had to submit to a court-martial, and though they cleared themselves Drake's opponents at Court who had gained the Queen's ear condemned him to inactivity. In vain she was urged to loose him on the Indies or suffer him to attack Cadiz so soon as the galleys were laid up for the winter. She would not listen, and the fruit of his splendid victory at Gravelines was allowed to rot away unplucked.

For the rest of the year Drake was busy winding up the accounts of the voyage and fighting his battle at Court. The inevitable result was that Philip's untiring persistence began again to be felt, and in the spring came fresh rumours of a great fleet gathering at Ferrol. The country grew alarmed, and so serious a panic seized Plymouth that the inhabitants began to abandon their homes. Drake at once hurried to his house in the town. There with his wife and household he quietly took up his residence and the panic was instantly allayed.[1] On the top of this new proof of the great admiral's power the

[1] *Lands. MSS.* lxv. 12.

Government received information that the King of Spain, having heard how Drake was in disgrace, was making overtures to him to enter his service. The intelligence came through Mr. Richard Drake of Esher, a kinsman and warm friend of Sir Francis, and thus may have been merely an invention to alarm the Government, so limitless was the Elizabethan capacity for intrigue. Still it had the desired effect. Richard Drake was an equerry in high favour, some of his kinsman's Spanish prisoners were in his custody, Philip had certainly attempted a similar stroke before, and Sir Francis was promptly given the command at Plymouth with orders to place the town in a state of defence and prepare with fireships and otherwise to resist any attempt by the Spaniards to retaliate on the western ports.[1]

It was but little consolation; a great expedition against Panama was in contemplation and Raleigh had obtained the command of it for himself. It was Drake's by right; it had been the dream of his life, and it was in the deepest mortification that he took up his small command. But even then his proud spirit had not felt the depth of its abasement. As it became known that the Spanish fleet was destined for Brittany, Drake asked to be allowed to attack it as it passed to the eastward, with the little squadron attached to his command. Even that was denied him. For since by the assassination of Henry the Third and the succession of the Huguenot king the centre of the great struggle had been shifted to France, the Queen was content to foster the Protestant cause by sending money and troops to the assistance of

[1] *S. P. Dom. Eliz.* ccxxxi. 94, April 1590; *Coke MSS., H. M. C. Rep.* XII. i. 13 and 14.

Henry of Navarre. Her navy was not permitted to complete its conquest of the seas, and all she would sanction was that a squadron should be maintained in the Atlantic to cruise for the Indian treasure-ships. At the last moment Raleigh's commission was revoked, his fleet was committed to Hawkins and Frobisher, and Drake had the last mortification of seeing it depart with his rival's flag floating over the *Revenge*.

That his self-esteem was deeply and even dangerously wounded is certain. For a while he permitted himself to contemplate an extravagant project for conquering the kingdom of Brazil for himself and Don Antonio, but the idea perished almost as soon as it was born.[1] Impatient and masterful as he had grown, his nature was too great and loyal to revolt or even to sulk under royal frowns. In all the ardour with which he had plundered and destroyed, he threw himself into the work of making Plymouth the strong naval port of his ideal. Not only did he carry out the works sanctioned by the Government, but almost entirely at his own expense he brought pure water into the town by a conduit many miles long and constructed flour-mills to provide the fleets with wholesome and sufficient biscuit, works for which his memory is honoured to this day by a yearly ceremony.

If Drake's princely gift to Plymouth had any other motive than a far-sighted patriotism, it was certainly nothing less noble than a desire to regain the favour of his capricious mistress. But he was not yet to be forgiven. When the Atlantic squadron came back from a successful but barren blockade of the Spanish trade-routes,

[1] *S. P. Dom. Eliz.* ccxxxi. 94.

Hawkins resigned and Frobisher was transferred to the narrow seas, but it was not Drake who took their place. The restless energy for which his work of coast-defence was not enough, he had to expend upon fitting out a privateer squadron and promoting a syndicate to assist him with funds. The command, as his favourite captain Fenner fell sick, he committed to the adventurous Earl of Cumberland with instructions to cruise off the Spanish coast on an inner line of blockade, and to intercept whatever escaped the Queen's ships at the Azores. For the Atlantic squadron, with Lord Thomas Howard in command, was again under orders for the Western Isles, and this was the immortal cruise when the *Revenge* was entrusted to Sir Richard Grenville. Who does not know the tale, and whose heart will not be moved when he tries to picture the injured Admiral as he heard how the ship he loved so well had fared? While he was eating out his heart between Plymouth and the great country-house at Buckland Abbey which Sir Richard Grenville had sold him, the pride of his life—the very embodiment as it seemed of his savage creed—had perished in the midst of the Spanish fleet fighting for a day and a night alone against a navy and dealing destruction around her to the last. Of the fifty-three sail she fought, four she sank, and in the end did not surrender till not a stick was standing, till every man, every pike, every barrel of powder was spent, and Sir Richard lay gasping out his life in impotent defiance. Even then she was not conquered. For true to her name, which to those old mariners meant so much, she went down with her prize-crew before she could be carried into port. A terrific storm the like of which no living man had seen was her knell, and about her grave were strewed along the shores of the Western Isles the wreckage of a hundred Spanish ships and three thousand Spanish corpses. "So it pleased them," Sir Walter Raleigh wrote, "to honour the burial of that renowned ship the *Revenge*, not suffering her to perish alone for the great honour she achieved in her lifetime."

It was a loss to Spain hardly less than that which the destruction of the Armada inflicted, but the tragedy was not yet at an end. For as the *Revenge* went down she seemed to summon Drake to his doom. There was felt to be something ominous in her loss. For so great a name had Drake won for her that she had been chosen as the model for the new ships that had been ordered after the defeat of the Armada, and this the most renowned of all her navy was the only vessel the Queen had lost in all the war. By her the shock was deeply felt, and ere she had time to recover from it there fell upon her the shadow of a danger greater than any she had yet encountered. The old Armada had failed partly because the invasion was attempted from a coast so dangerous and so hard to reach as that of Flanders, and partly from the faulty design of the unwieldy Spanish ships. But now Philip with splendid patience was constructing a new navy on French and English lines, and in Brittany he had secured against his enemy an advanced base which was untainted with any of the vices of Dunkirk. The danger was plain, great, and imminent, and in the hour of her trouble the Queen remembered the man who had so often filled her purse, and had never lost her a ship. For all his long disgrace he was the one figure that stood out in proportions large enough to grapple with the peril; and as the summer wore away the merchants of Cadiz began to whisper in alarm that the Dragon was to be unchained once more.[1]

In the autumn Drake was summoned to Court, and once there he carried all before him. Raleigh was in disgrace, Frobisher and Lord Thomas Howard paled again beside the new-risen sun, and ere long the realm was all astir with the growth of a kingly project. In February Parliament was summoned. As soon as it met it was asked for a grant of two hundred thousand pounds, and told that if it were voted the Queen was resolved to send Drake to sea to encounter the Spaniards with a great navy. So unprecedented was the sum, however, that the bill met with serious opposition. The Admiral, who worthily sat for King Arthur's Castle of Tintagell, supported the Government with all the weight of his influence and eloquence. He served on all the public committees except that to which the bill for the suppression of the Puritans was referred, and on all others where he had a special or local knowledge. For the subsidy bill he did his best with a speech in which he graphically depicted the horrors of a Spanish conquest, and after debates which lasted eleven days—a very lengthy discussion as things went then—the bill was passed and Drake was at once plunged into a mass of business relating to naval affairs and coast-

[1] *S. P. Ireland*, "Advice from Spain," June 16th, 1592. *H. M. C. Rep.* vii. 524a, Ph. Gawdy to his brother, December 8th, no year, but certainly 1592 from internal evidence. Cf. Birch, *Memoirs*, i. 90.

defence. But it was all too little to allay the pent-up impatience of his long disgrace. He could not wait a year, perhaps two, till Elizabeth and Philip had fully armed. He was burning to act. Panama, the virgin treasure-house of his desire, was still unsacked. Raleigh had nearly supplanted him, and from the first moment he came to Court he began tempting his mistress again. In January, 1593, he presented her with an account of his first romantic voyage, when from the little grove beside Pizarro's road he had seen and desired. But as though some presentiment of the end hung over her she could not bring herself to let him leave her side, nor was it till the summer of 1594 that he prevailed.

Then at last, so magnificent was Philip's deliberation, Drake persuaded Elizabeth how easy it would be to make a dash over the Atlantic for Nombre de Dios, throw a few troops across the Isthmus to sack Panama, and be home again with all the wealth of Peru before the new Armada could sail. It was of course to be quite a private adventure, but the Queen was to provide two-thirds of the capital. The rest was to be underwritten by Drake and Hawkins. She was also to lend six ships to escort the transports, and as the success of the undertaking depended chiefly upon the land forces, Drake was authorised to proceed to Holland and seek the consent of the States to the recall of the English veterans in their service.[1]

His mission was a failure, and it was but the first of a succession of disappointments. Voice after voice

[1] Umpton's despatch, *Murdin*, 733, March 20th, 1596. Burleigh to Sir Robert Cecil, *S. P. Dom. Eliz.*, July 21st, 1594, seems also to refer to this visit.

seemed to warn him of his doom, but nothing could shake his self-confidence. All the winter, from his great house by the Dowgate, which had sheltered so many of the most famous of English warriors, he struggled with the difficulties that beset his preparations, and as the ring of his toil echoed in the ears of Philip's recruits, they deserted by hundreds, for very terror of his coming. Lisbon itself was almost stripped of inhabitants. Ten thousand householders fled with their families in a panic, and those that remained sent away their wives and goods. The English spies declared that Drake's name was more feared in Spain than ever was Talbot's in France, and from the highest to the lowest there was no talk but of the coming of Drake.[1] At home the magic of his name had no less power. Volunteers flocked to his flag in such numbers that he hardly had to press a man, but for all their eagerness they were a sorry substitute for the tried campaigners of the Low Countries. Nor had he a Norreys to command and discipline. The Chickens of Mars had quarrelled with him as they quarrelled with every one else, and Drake was a man even harder to reconcile than those turbulent knights. Another family, only a degree less renowned, supplied their place. The commission of colonel-general was given to Sir Thomas Baskerville, the hero of Sluys and Bergen-op-Zoom. The first of the famous captains that make a halo for the name of Sir Francis Vere, at the head of but one hundred men he already had taught eight companies of the Spanish Old Legion that they were no longer the finest infantry in the

[1] Report from Spain, *S. P. D. Cal.*, June 9th, 1595. Halliday to Burleigh, *S. P. Spain*, xxx. 5a, March 16th, 1595.

world, and during a truce Parma himself had embraced him, proclaiming that no braver soldier served a Christian prince. On his staff were his brothers Arnold and Nicholas, and Sir Nicholas Clifford was his lieutenant—a band worthy to serve under the flag of Drake. But as the enterprise grew ripe, the Queen, always half frightened at the ungovernable energy of her favourite engine, again lost heart, and not content with having set the caution of Hawkins to drag upon it, in May she refused to let them go at all. The old fatal hesitation had once more seized her. For two months they were compelled to lie idle consuming their resources and losing their men, till early in July she had again plucked up heart, and they were again told they might go. With twelve chartered merchantmen and the six navy ships they at once hurried out of the Thames to join the rest of their squadron at Plymouth, and while Hawkins toiled to complete the ships for sea, Drake scoured the West Country for supplies to set the force on its legs again. By the end of the month they hoped to be ready, when into the midst of their final preparations broke the news that Penzance was in flames. Four Spanish galleys, supported by a fleet of forty sail, had suddenly slipped across from Brest to retaliate for Drake's insults to the Spanish coast, and while Baskerville flew westward to rally the county levies, Drake in a fury dashed out of Plymouth with the ships that were ready. It was too late. The Spaniards had heard their terror was at hand, and having destroyed a few fishing villages, hurried off faster than they came.

Stung by the blow the Spaniards had so smartly given in their very faces, the admirals were now more eager than ever for their sailing orders. It was the 11th before the expected packet came, and when they opened it, it was to be astounded with an entire change of plan. Instead of making a sudden invasion of the Darien province, they were ordered first to cruise off the Spanish coast for intelligence, then to capture the Plate fleet, and finally, before receiving their route, they were told they must promise to be back in six months, in order to serve against the new Armada. In despair the admirals protested that they could not go cruising or fight a naval action with a fleet of transports, and as for promising to be back by a certain day, it was but tempting Providence. They were ready to obey the Queen's orders, they said, but they must have time to disband the troops and reorganise the expedition. Elizabeth was furious at their perverseness and disloyalty, as she was pleased to consider it; nor is it possible to say how long the dead-lock would have continued had not news of a disabled treasure-ship of enormous value lying at Puerto Rico suddenly tempted the Queen into reason.

It was already the end of the month. There was no time to do all that had been undone, and with but twenty-seven sail and only two thousand five hundred men they started on the ill-omened voyage. A Spanish fleet was known to be out, and so nervous was the Government about Ireland, that Drake in spite of his protests had to open his campaign with a reconnaissance on the coast of Spain. On his way he was chased by twenty strange sail, but he showed them a clean pair of heels, and a few days later, having ascertained from a frigate of the Earl of Cumberland's that the

Spanish fleet was going home, he continued his voyage for the Canaries. Drake and Baskerville, in consequence of the course forced upon them by the Queen's instructions, had found that it would be necessary to land there to water and refresh the soldiers. Hawkins, who less understood the necessities of a military expedition, violently opposed Drake's proposal, and the result was a painful scene between the two old friends at the council table. The soldiers, however, persuaded them to dine together on Hawkins's flagship next day, and the old admiral was brought round to his kinsman's view. Four weeks out, therefore, the fleet anchored under the guns of Las Palmas in the Grand Canary, but it was only for Drake to get his first hint that the days for his daring raids were over. Ever since it had been known in Spain that he was again in favour, Philip had been busy reinforcing and fortifying his colonial ports, and Las Palmas had not been forgotten. To surprise it there and then might have been possible; but Drake found the surf too high for a landing and drew off after a three hours' search in his barge for a practicable beach. Baskerville offered to take the place in four days by a regular operation, but Drake would not wait. Too much time had already been lost, and having watered at another part of the island the voyage was next day continued across the Atlantic.

Two days from their rendezvous at Guadeloupe a gale broke up the fleet, but in spite of it Drake managed to get all his squadron safely anchored behind Ste. Marie Galante. Thence he proceeded to his well-known anchorage, and there next day Hawkins joined him prostrate with misfortune. A small ship of his squadron had been captured by five Spanish frigates bearing for San Juan de Puerto Rico. The whole gravity of the disaster at once flashed on Drake. He divined immediately that the enemy's squadron must be the ships which had been sent to embark the treasure from the disabled galleon, and but for the unfortunate straying of the lost tender he knew the whole of them must have fallen into his lap. Now not only had they escaped, but from their prisoners the Spaniards would torture the secret of his destination. He was for weighing on the spot in pursuit, but his colleague had lost his nerve. The old admiral was ill and anxious, and querulously insisted on first cleaning the ships and getting the big guns up from the holds into their places, that the fleet might be ready for anything. Then in pity for his old friend's condition Drake did what he is never recorded to have done before,—he gave way.

Four days they lay getting the fleet into fighting trim, and, when at last they sailed, from Guadeloupe to Tortola not a trace of the enemy could be seen. It was clear they had reached San Juan to reinforce and warn the garrison, and unless it could be thrown off its guard again Drake knew that success was almost impossible. Then in the hour of need the spirit of his youth came once more strong upon him; the fleet seemed to grow no more cumbrous in his hand than a privateer schooner; and falling back on the bewildering tactics of his buccaneering days he made it disappear from the seas. In vain the Spaniards watched for his coming. Just as twenty years ago his two tiny craft lay lost in the creeks of the Darien Gulf, so now into the still recesses of the Virgin Islands in roadsteads unknown he had led his whole fleet where none could guess of its existence, and there for some days they lay drilling the soldiers and refreshing them ashore, while Drake in his barge surveyed for an outlet.

But for old John Hawkins the change brought no relief. Grieving over his misfortune he was falling deeper and deeper into the clutch of his sickness. He was sinking fast when Drake having discovered a secret channel to the southward sallied out from his hiding-place, and slipping in behind the Spanish scouts appeared unannounced before Puerto Rico.

Sounding as he went, he brought up the fleet to the astonishment of the Spaniards in a road where no ships had ever been seen to anchor, and there the dying admiral breathed his last. His death was perhaps no loss to the expedition. It had suffered already severely from the caution which grew upon him with age and failing health. But if as a fighting admiral his reputation was lower than that of others, yet no one could forget that it was to him that England owed all that was good in her navy, and the gloom which the fine old seaman's death cast over the fleet was still to be deepened. As Drake sat at supper that evening on board his flag-ship the *Defiance* discussing the forthcoming attack with his officers, a round shot crashed through the cabin. Drake's chair was smashed under him; Sir Nicholas Clifford, who alone had supported Hawkins in his fatal counsel, was killed on the spot; and a young officer named Brute Brown, to whom Drake was particularly attached, was mortally wounded. It is no wonder that under the shadow of these losses the fleet was removed out of range, and that nothing further was attempted that night. On the morrow, however, Drake again astounded the Spaniards by bringing up his fleet in another unknown anchorage close to the town and yet masked from its guns. He spent the day in his barge seeking a weak point to attack, but the result of his reconnaissance was only to show how hard a task he had before him, and he resolved that night to burn the frigates which had come for the treasure, that it might not escape him while the town was being reduced. In person he led the boats to the mouth of the harbour, and having shown each its station he retired. A desperate fight ensued. Again and again the vessels were fired, and as often they were extinguished. At last one was fairly in flames, but by its light the garrison, reinforced by the newly arrived crews, poured in such a murderous fire from the shore that the English were compelled to retire with heavy loss. The failure was disastrous; but so far from discouraging Drake, it stung him into perhaps the most daring resolution of his adventurous life. Where the boats had failed he knew the ships could succeed, and determined to wipe out his defeat he made up his mind to carry the whole of his warships right inside the port, and crush the enemy with an overwhelming fire. To carry out this unprecedented stroke, during the following morning he worked the fleet to windward and in the afternoon ran down free for the mouth of the harbour. Already it had been partially blocked by sinking the great treasure-galleon, and as the English approached, the Spaniards were seen to scuttle three more vessels at her side. To continue his course was to wreck the whole fleet, and Drake was compelled to abandon his splendid attack and bear up again for his anchorage. Unwilling, however, to abandon the enterprise altogether, on the morrow he made a new reconnaissance; but so greatly had the Spaniards been able to strengthen the fortifications since the arrival of the frigates, that he was convinced the capture of the place by land operations was beyond the power of his weakened force. In vain some of the soldiers urged a new attempt. Bold as Drake was, none knew so well as he when daring merged into folly. In his gasconading way he told them he could bring them to a score of places richer and more easy to take, and years ago before irresolution had ruined England's chance he could have made good his boast. The soldiers still believed he could perform his promises. And so as the evening closed in Hawkins and Clifford were solemnly committed to the sea, and under cover of the night, for the first time in his life, Drake bore away beaten.

Perhaps it had been better if he had stayed and fought it out. But as his end drew near, the scenes of his youth seemed to call him with an irresistible voice. As he made his boast to the soldiers there was in his mind a day long ago, when he had had his first sight of the fabled Indies and under Captain Lovell had lost his all. His life was waning: that day was still unavenged; and across the Caribbean Sea Drake led his squadron to La Hacha. This time he did not even permit the fleet to appear in sight of the threatened port. The only effect of his failure seemed to be to produce a distrust of all but the daring shifts on which his reputation had been founded; and as in his first great exploit he had crept with his three boats into Nombre de Dios, so now from Cape de la Vela he sent his flotilla to steal along the coast and surprise La Hacha. But it was no case for surprise. On his arrival the following day with the fleet, he found his troops in unopposed possession of a town deserted and stripped bare. He seized the neighbouring pearl-fishery, but that was bare too, and for a week no offer of ransom came. At last he agreed to accept thirty thousand ducats, but when at the end of another week the pearls in which payment was to be made arrived, they were found to be unconscionably over-valued. With a magnanimity almost quixotic, Drake refused to touch a pebble, and instead of keeping what he had got and demanding more, he chivalrously gave the envoys four hours to clear themselves and their treasure. Not another ducat did he get. He had yet to learn a new lesson, and find how little a roving force like his can do against a great empire resolute not to ransom open towns. The governor had arrived, and refused so much as to discuss the question. On the morrow Baskerville burnt an inland village. It had no effect, and next day Drake fired the town, as well as the *rancheria* of the pearl-fishers and all their boats. Thus he took his long-deferred revenge upon La Hacha; but, true to his simple creed, he would not suffer a hand to be laid upon the church, and with a touch of that gentleness that gilds with knightly grace his most savage devastations, he spared from the flames the house of a lady who begged his mercy. At Santa Marta, farther to the westward, the same scene was played; and on Christmas Day, with the smoke of the burning settlement rising behind him, he bore away for Nombre de Dios.

Here, too, the town was found deserted by its inhabitants, and was occupied almost without resistance.

At the first sign of their approach the governor had fallen back to Panama, and without losing an hour Baskerville started in pursuit with seven hundred and fifty men. It was all that sickness and casualties had left available for the service, for it would seem to have been Drake's intention to ascend the Chagres river with another column. For three days, however, he remained where he was, to search for buried treasure and to fire the town. This done he prepared to weigh, but ere the anchors were up there came flying through the blackened ruins of the Panama gate a message of disaster. Baskerville was in full retreat on the ships, hurled back from impassable intrenchments.

Then it was that the undaunted heart began to wax cold. The jovial face grew sombre. The cheery smile, to which his men had ever been accustomed to look for light in the darkest hours, had faded, and failure began to haunt him, as he recognised how the terror of his name had changed the Indies. The seas were deserted, the ports bristled with guns, and feverish wakefulness had supplanted the old dreamy security. Yet not a word of doubt was suffered to pass his lips. "It matters not, man," he would cry to any croaker that repined, "God hath many things in store for us; and I know many means to do Her Majesty good service, and to make us rich: for we must have gold before we see England." He called his council of war together, and showed them on the map Truxillo, the famous port of Honduras, and the El Dorado, where the golden towns lay clustered around the Lake of Nicaragua. He asked which they would have, and stout-hearted Sir Thomas Baskerville cried "Both!" So it was resolved without more ado.

All the Spanish shipping in the port was given to the flames, and thus, still marking with fire the road to his grave, and abandoning himself to the old adventurous dreams of his boyhood, he led his desperate treasure-seekers across the Mosquito Gulf.

It was a quest as wild as any his buoyant youth had dared. He knew no more of what lay before him than what the ill-drawn maps revealed, nor had he now the high spirits that were wont to make a sport of danger. Damped with failure, and in cold blood, he was bent on saving his reputation or dying with it. The very hand of God seemed to wave him back as he struggled westward against contrary winds, until so foul and boisterous grew the weather that, driven at last into the depth of the Gulf, he had to take shelter behind a desert island called Escudo de Veragua. In all the Indies no spot displayed in fouler guise the black side of the western paradise. The radiant tropic growth of flower and palm invited the sick to seek recovery upon the putrid soil beneath, and as the boats bore them into the fairy creeks loathsome reptiles started from a slime that reeked of death. But Drake would not give way. Day after day he clung under the lee of the deadly shore hoping each hour for a wind to carry him on. With his eyes still bent forward he kept the men busy setting up the pinnaces which would be required to ascend the San Juan river, that the delay might not be lost time. But day after day the wind continued foul, and with each returning dawn new victims sank in the poisonous air. Drake himself was stricken with dysentery, but still he strove against his fate till he had no strength to leave his cabin. Nor was it till his sickness had imprisoned him a week that he consented to weigh and let God take him where He would. But it was now too late. He had drunk too deep of the island's breath. He had defied his fate too long. As the fleet sped back eastward the pestilence ran riot through the ships, and the Admiral lay still and conquered in his bed. Broken in spirit he could not shake off the disease, and when, after a week's tempestuous voyage, the fleet anchored off Puerto Bello, he was lying speechless at death's door. But it was not thus a soul so stubborn could pass away. It was the 28th of January 1596, and as the dawn of his last day broke a delirium seized him. He rose from his bed and clothed himself, calling like a dying Viking for his arms, and raving in words none cared to record. Yet we can hear him railing at traitors who had stolen his life with poison, and see in his last frenzy the origin of those envenomed rumours that whispered of foul play. His fury past they led him back to bed, and there at last as quiet as a sleeping child the sea-king died.

In the fine roadstead of the newly-founded port the fleet cast anchor, and as the news spread from ship to ship the first desire of all was to hurry home as best they could like sheep that have lost their shepherd. Of continuing the venture there was not a thought, save to seize the half-built settlement for a burnt sacrifice to grace the burial of the dead commander. On the morrow the last rites were performed. Enclosed in a leaden coffin the body was carried a league to sea, and there in sight of the spot where his first victory had been celebrated, amidst a lament of trumpets and the thunder of cannon, the sea received her own again. At his side were sunk two of his ships for which there was no longer need and all his last-taken prizes, and for a pall he had the smoke of the latest fort which his life-long enemy had raised against him. So the fleet went its way, and left him lying like a Viking dead and alone amidst his trophies on the scene of his earliest triumph and his last defeat.

At home, while the weed waved over that silent resting-place, every dockyard was noisy with hum and clatter of shipwrights as the great fleet grew and gathered, and while the Spanish Indies made high holiday over the Dragon's fall, England in busy anticipation watched vainly westward. The arm for which she was forging the mighty weapon was never to wield it. To such a supreme effort Drake had toiled to spur her through long years of labour with his life in his hand, too often with a rope round his neck; and now on the eve of accomplishing his life's work he was dead like a sick girl of an inglorious death. Others were to reap where he had sowed, and hands unfit were to prove the hero's sword. In their eager grip it bit deep and hard, the giant reeled with a gaping wound in his side, but then the sinews of his assailants failed and he arose again dogged and huge and terrible still.

THE END